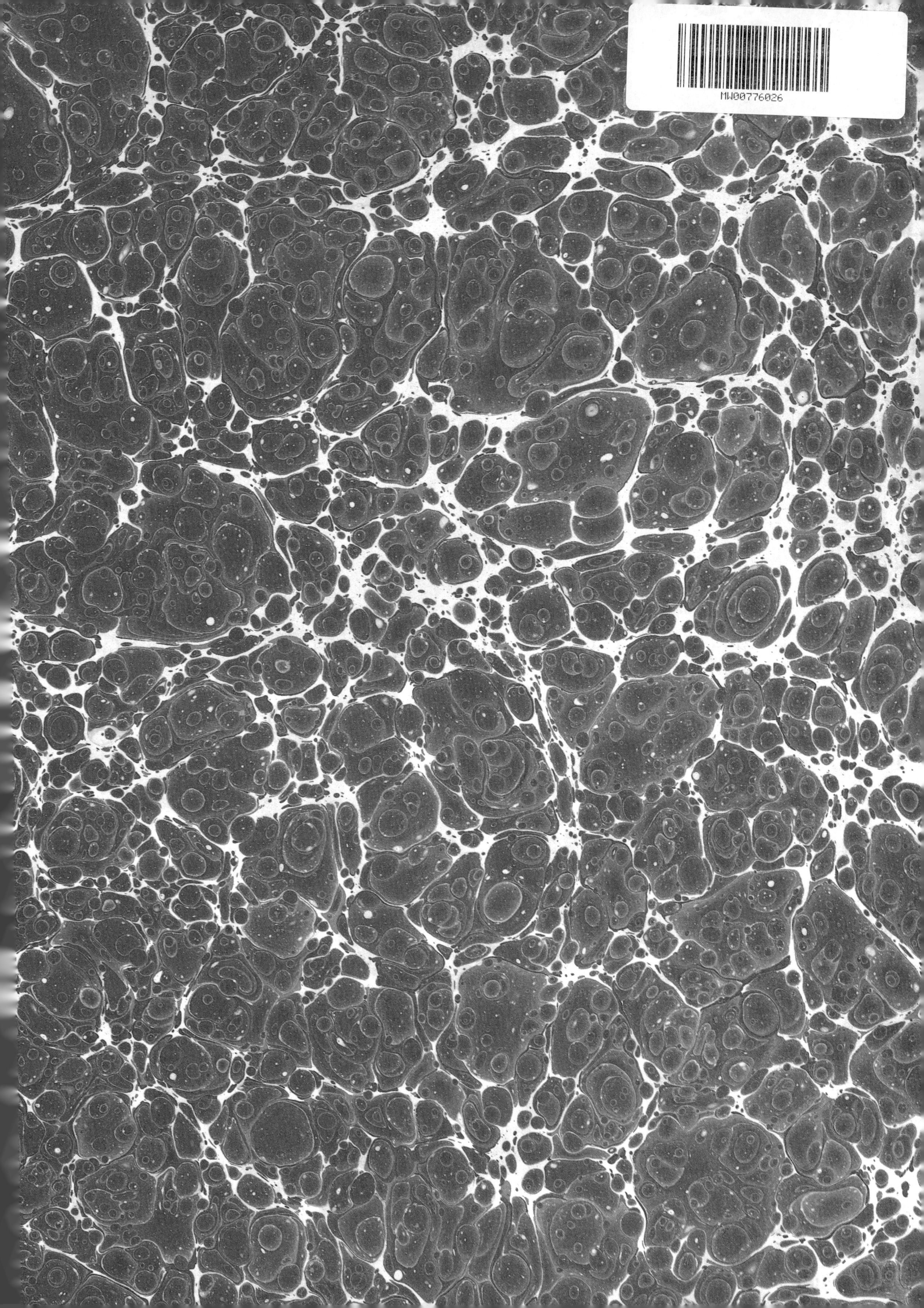
MW00776026

The Cavalry Regiments of Frederick the Great 1756-1763

Günter Dorn

Joachim Engelmann

The Cavalry Regiments of Frederick the Great 1756-1763

Günter Dorn　　Joachim Engelmann

1469 Morstein Road, West Chester, Pennsylvania 19380

BIBLIOGRAPHY

Geschichte der Preussischen Armee (History of the Prussian Army), by Curt Jany, 2nd Edition, Vol. I-IV, Biblio-Verlag, Osnabrück, 1967

Stammliste aller Regimenter und Corps der Königlich Preussischen Armee für das Jahr 1806 (Register of all Regiments of the Royal Prussian Army for the Year 1806), Reprint by Biblio-Verlag, 1975

Soldatisches Führertum (Military Leadership), by K. von Priesdorff, Vol. I-III, Hamburg, no date

Die geistige Kultur des preussischen Offiziers (The Spiritual Culture of the Prussian Officer), by Friedrich-Karl Tharau, von Hase & Köhler Verlag, Mainz, 1968

Geschichte der Küniglich Preussischen Kürassiere und Dragoner 1619 bis 1870 (History of the Royal Prussian Cuirassiers and Dragoons, 1619 to 1870), by Alt, Schropp'sche Hof-Landkarten-Handlung, Berlin, 1870

Die Preussischen Husaren—Von den ltesten Zeiten bis zur Gegenwart (The Prussian Hussars—from the Oldest Times to the Present), edited by Fr. Krippenstapel and R. Kntel, Berlin, 1883, reprinted by Jürgen Olmes, Krefeld, 1970

Die Kavallerie des Deutschen Reiches—Derselben Entstehung, Entwicklung und Geschichte (The Cavalry of the German Empire—its Origin, Development and History), edited by R. von Haber, Weimar, 1877, reprinted by Jürgen Olmes, Krefeld, 1977

Die Altpreussische Armee 1714 bis 1806 und ihre Militärkirchenbücher (The Old Prussian Army and its Military Hymnals), by Alexander von Lyncker, Verlag Degener, Neustadt/Aisch, 1980

Geschichte der Reiterattacken (History of Cavalry Attacks), edited by Carl Bleibtreu, Verlag A. Schall, Berlin, no date

Die Ritter des Ordens Pour-le-Merite (The Kinghts of the Order of Pour-le-merite), edited by Gustav Lehmann, Vol. 1, Verlag Mittler & Sohn, Berlin, 1913

Die Uniformen der Preussischen Kavallerie, Husaren und Lanzenreiter 1753 bis 1786 (The Uniforms of the Prussian Cavalry, Hussars and Lancers, 1753 to 1786), by Hans Bleckwenn and F.-G. Melzner, Part III, Vol. 4, Biblio-Verlag, Osnabrück, 1979

Translated from German by Dr. Edward Force.

Library of Congress Catalog Number: 88-64006.

Printed in the United States of America.
ISBN: 0-88740-164-3
Published by Schiffer Publishing, Ltd.
1469 Morstein Road, West Chester, Pennsylvania 19380

This book may be purchased from the publisher.
Please include $2.00 postage.
Try your bookstore first.

CONTENTS

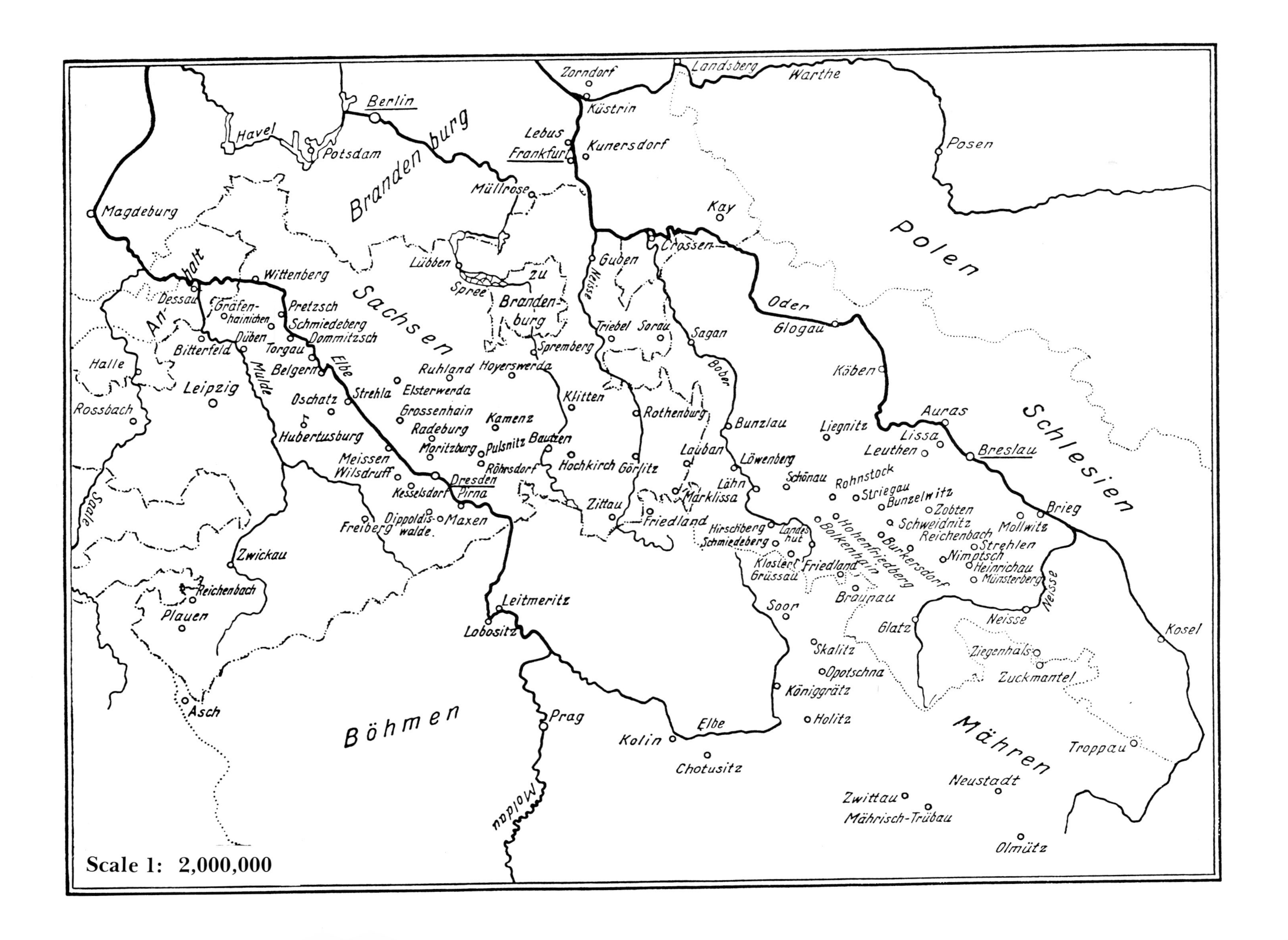

Scale 1: 2,000,000

PREFACE

This splendidly prepared historical uniform volume with lavish full-page illustrations,

THE CAVALRY REGIMENTS OF FREDERICK THE GREAT

is an outstanding work that will certainly be accepted with enthusiasm by wide circles.

It is not only a valuable reference book for persons interested in Prussian military history, but will, above all, be greeted with joy by old cavalrymen and riders too.

Until well into the Nineteenth Century the cavalry was the weapon that won battles, comparable to the armored units of World War II.

The training, equipping, arming and preparation of his cavalry was especially close to the heart of the Great King.

Two outstanding cavalry leaders, Generals von Seydlitz and Zieten, stood loyally at his side. Their motto was forward thinking, forward looking, and forward riding. They made sure that the spirit of the rider and the knight were at home in the cavalry regiments, an important basis for harmony between rider and horse.

May this splendid uniform volume fine widespread acceptance!

Horst Niemack
Major General, retired
President of the Order of
the Knight's Cross Bearers and the RDS

FOREWORD

Frederick the Great built up the Prussian Cavalry, in view of the numerical and qualitative superiority of its Austrian counterpart, which in many ways was not only an enemy but an example too, at least as strongly as the infantry with its excellent foundation. He expanded the cuirassiers, in existence since the time of the Electors, by one regiment, the dragoons, who dated back to Friedrich Wilhelm, to almost twice their numbers, and multiplied the hussars tenfold. In all, he created fourteen regiments, two-fifths of their totals in 1786. He can be regarded above all as the creator of the hussars, for which he even brought in Hungarian, Balkan and Polish members. The origins of the later Uhlans also came about in his time. The cuirassiers were the oldest type of mounted force, being called "Regiments on horseback" since 1623, known informally as cuirassiers since 1758 and officially since 1786. The first dragoons came into being in 1631, the first hussars in 1721.

The cuirassiers were the heavy battle cavalry, the dragoons originally mounted marksmen and yet capable of making cavalry attacks, the hussars actually just light cavalry for reconnaissance, raiding and securing. After the cavalry's failure at Mollwitz in 1741, the King devoted particular care to these troops in an ongoing series of tactical advice, instructions and requirements, particularly in 1743. At Hohenfriedberg in 1745 one regiment completed the victory. Before the Seven Years' War the King trained the cavalry to be "indomitable", banned artificial maneuvers and carbine salvos, and ordered ruthless closed attacks "like a wall" at the right moment, with repeated gathering and charging. At Rossbach in 1757 the cavalry won the battle alone. By developing its varied special tasks, it grew together into a battle cavalry that could move from the flanks—in follow-up moves as well—and win or contribute to the victory through determination, quickness and attacking power.

The cavalry always had an air of aptitude, nobility and daring about it, as well as chivalry, which symbolized a whole epoch of culture. Since Prince Eugene and Marlborough, their numbers in the Seventeenth and Eighteenth Centuries always amounted to the "smaller half" of the army, while their strength by the end of the Eighteenth Century, as a result of the expansion of the infantry, shrank to about one seventh. It was the century of the cavalry! From their 18,000 men in May of 1740 to their 37,500 man in 1763—the strongest of their time—the Prussian Cavalry achieved their great contribution to Prussia's survival. They were, as Tempelhoff reports, "filled with a lively sense of honor and a spirit of bravery that taught them to bear the greatest burdens with patience and steadfastness and gave them life and energy in the most dangerous situations".

Joachim Engelmann

INTRODUCTION

Cuirassiers have been known since 1481 and are the oldest type of cavalry. The rider, formerly in full armor, then wore only a breastplate, originally of leather (French: cuir), which in the end was only worn in parades. They came into being as "regiments on horseback", and later were called "heavy cavalry"; as of 1666 they were permanent regiments, whose development was ended as of 1718. The use of the designation "cuirassier" in commands began in Prussia in July and August of 1741, was already widespread by 1742, and was used unofficially since 1758 and officiallysince 1786. The cuirassiers formed the battle cavalry, used for preliminary fighting on the wings and for the attack at the right moment, usually against a wing. In the Eighteenth Century they usually fought in the first encounter, as long as horse and rider were bigger. The weapons they bore were the carbine and the "Pallasch", a broader and heavier sword than the "Degen" and with a back. They were regarded as a splendid elite troop, used on an average of 55% in eighteen battles.

The regiments had five squadrons, whose man-power was increased again and again. From a specified strength of 132 privates in 1740 they increased to 156 in 1767 and to 168 soldiers in 1757. Because of losses and the war situation, they sank after 1758 to the level of 1744, with 144 privates, with undiminished numbers of officers, and remained unchanged until 1762, when they were expanded to 174 soldiers. The ups and downs corresponded the war situation and the availability of replacements. Every squadron had two trumpeters, two smiths and one medic. In 1740 the regiment numbered 834 soldiers, 884 in 1758, and in 1762 it added up to 1000 soldiers. Surprisingly enough, the King still thought in terms of expansion at the end of the Seven Years' War. Only C.R. 13, the Guard Corps, consisted of one squadron in 1740, but of three as of October 1756. Considering losses and departures, we must regard a squadron as numbering 130 to 150 men in practice. The regiment was—disregarding mobility and attacking power—basically just as strong as a decimated infantry regiment.

The dragoons emerged in the Thirty Years' War as "infantry set on horseback" and developed into a second genre of cavalry as they fulfilled their original double role. France had developed them under various names as mounted musketeers. Prince Eugene was called the "Father of the Dragoons", Derfflinger brought them to Brandenburg in 1687, where they were established as regiments from 1689 to 1744, with the help of refugees. Here too, the weapon gave the troop its name. The "dragon" was a shortened carbine with a dragon's head at its mouth. For a long time firing from horseback was done in the platoon manner; it was not easy to teach that. As a rule, they formed the second encounter behind the cuirassiers. When they performed particularly well, they were accorded the "Riders' March" or were even promoted to "Reuter", as many regiments were by King Friedrich Wilhelm I after the Pomeranian campaign of 1715. This made their place in the ranks clear. In eighteen battles they were involved on an average of 38%.

Like the cuirassiers, the regiments had five squadrons, which were divided into two companies longer than the cuirassiers were. Only Dragoon Regiments 5 and 6 had ten squadrons. The man-power of the squadrons corresponded to that of the cuirassiers, with 132 in 1740, 144 in 1744, 156 in 1756 and 168 in 1757. But the decrease of 1758 did not affect the dragoons; their optimal strength remained unchanged until 1762, when it rose to 174 men. Every squadron had a smith and three drummers, which were abolished only in 1771, a mark of their origin in the infantry, as were the designs on their bullet-pouches. The total strength of a regiment was somewhat higher than that of the cuirassiers: 837 men in 1740, 892 in 1756, and 907 as of 1757, until they equaled the cuirassiers with 1000 men in 1762. The lower staff of the regiment consisted of one quartermaster, chaplain, auditor, regimental medic with five medics, one preparer, saddler and provost, plus one drummer and four oboists as in the infantry.

The qualities and accomplishments of the regiments varied; some were quite the equal of many a cuirassier regiment.

The hussars, regular troops only as of 1688, were the youngest genre of the cavalry, apart from the uhlans. After having proved their worth by 1712, they were instituted in 1721 and were ranked with the dragoons until 1735. Friedrich II increased their ranks from nine by nearly ninety squadrons, making a hundred in all. They were volunteers with an atypical life style that differed from the usual military ways of most regiments, being light cavalry, intended for patrols, raids, surprise attacks, pursuit, and securing of transport. In spite of being formed in regiments, they maintained the unaffiliated spirit of free corps with their always bold, often astounding maneuvers, eager for action, acting quickly, never giving in, independent. It is remarkable how well the Hungarian and Polish-Slavic influences blended with the Prussian. The King developed their use in battle in the third encounter, as flank protection and a reserve. Since September 24, 1741 the hussar regiments had ten squadrons, with lesser strength than the cuirassiers and dragoons. In 1743 the squadrons consisted of 102 privates, one trumpeter, smith and medic. Only in 1757 did they add up to 114 soldiers, with the exception of H.R. 5 and 7; in 1761 they totaled 136 men, in 1762 they reached 150.

The hussar regiments had an optimal strength of 1166 in 1743. Their lower staff consisted of a quarter-master, regimental medic, two gunsmiths and two bootmakers. From 1743 on they no longer carried flags. In 1757 their numbers grew to 1301 soldiers, which was not attained in H.R. 3, 5, 6, 7 and 8. Only in 1759 did H.R. 3, 6 and 8 surpass 1300 soldiers. In 1762 they had 1681 men. In eighteen battles they were used on an average of 40%. The high point of their success was between 1792 and 1815.

Preserving the cavalry spirit in the field and the garrison, energetic use of man and horse in war and peace, tactical training of cavalry officers and years of war experience made the Prussian Cavalry capable of their accomplishments.

CAVALRY REGIMENTS BY PROVINCE
1740- 1786

ark Brandenburg

.. 2	founded 666, garrison: Kyritz etc. cadre: recruitment, replacements: Brandenburg
..7	founded 1688, garrison: Salzwedel etc. cadre: Prussia, replacements: Brandenburg
.. 10	founded 1691, garrison: Berlin cadre: Magdeburg, replacements: Brandenburg
.. 11	founded 1691, garrison: Rathenow etc. cadre: Mark, replacements: Brandenburg
.. 13	founded 1740, garrison: Potsdam, Berlin cadre: recruitment, replacements: selection from regiments
..1	founded 1689, garrison: Schwedt etc. cadre: Ansbach, Rhineland, replacements: Pomeranian
..3	founded 1704, garrison: Friedeberg/N. cadre: Brandenburg, Prussia, replacements: Brandenburg
..4	founded 1704, garrison: Landsberg/W. cadre: Brandenburg, Prussia, replacements: Brandenburg
..2	founded 1730, garrison: Berlin cadre: recruitment, replacements: I.R. 19 & 25

merania

. 5	founded 1683, garrison: Belgard etc. cadre: Westphalia, replacements: Pomerania
.. 5	founded 1717, garrison: Pasewalk etc. cadre: transfers; replacement: Pomerania
.. 12	founded 1742, garrison: Greifenberg cadre: Württemberg, replacements: Pomerania
.. 8	founded 1743, garrison: Stolp etc. cadre: Hungary, Magdeburg, replacements: cavalry regiments

st Prussia

.. 6	founded 1717, garrison: Königsberg cadre: Saxony, replacements: Prussia
. 7	founded 1717, garrison : Tilsit cadre: Saxony, replacements: Prussia
. 8	founded 1744, garrison: Insterburg cadre: Prussia, replacements: Prussia
. 10	founded 1741, garrison: Osterode etc. cadre: Pomerania, replacements: Prussia
.. 5	founded 1741, garrison: Goldap etc. cadre: Silesia, Prussia, replacements: D.R. 6 & 8
.. 9	founded 1745, garrison: Goldap etc. cadre: Bosnia, replacements: recruitment and I.R. 53

West Prussia

D.R. 9	founded 1741, garrison: Riesenburg etc. cadre: Pomerania, replacements: Pomerania, Prussia
H.R. 7	founded 1743, garrison: Schneidemühl etc. cadre: Hungary, replacements: D.R. 12
H.R. 10	founded 1773, garrison: Soldau etc. cadre: transfers, replacements: West Prussian

Silesia

C.R. 1	founded 1666, garrison: Breslau cadre: recruitment, replacements: Silesia
C.R. 4	founded 1674, garrison: Neustadt etc. cadre: Brandenburg D.R., replacements: Silesia
C.R. 8	founded 1689, garrison: Ohlau etc. cadre: Mecklenburg, replacements: Silesia
C.R. 9	founded 1691, garrison: Oppeln etc. cadre: Brandenburg, replacements: Prussia, Silesia
C.R. 12	founded 1704, garrison: Ratibor etc. cadre: Prussian D.R., replacements: Prussia, Silesia
D.R. 2	founded 1689, garrison: Lüben etc. cadre: Ansbach, Rhineland, replacements: Pomerania, Silesia
D.R. 11	founded 1741, garrison: Sagan etc. cadre: Silesia, replacements: Silesia
H.R. 1	founded 1721, garrison: Herrnstadt etc. cadre: Hungary, replacements: D.R. 2 & 11
H.R. 3	founded 1740, garrison: Kreuzburg etc. cadre: Hungary, replacements: C.R. 1, 9, later 8
H.R. 4	founded 1741, garrison: Oels etc. cadre: Polish, replacements: C.R. 1 & 8, D.R. 2
H.R. 6	founded 1741, garrison: Peiskretscham etc. cadre: recruitment, replacements: C.R. 4, 9, 12

Magdeburg/Halberstadt

C.R. 3	founded 1672, garrison: Schönbeck etc. cadre: Brandenburg, replacements: Magdeburg
C.R. 6	founded 1689, garrison: Aschersleben etc. cadre: Westphalia, replacements: Magdeburg

Within the provincial divisions, cuirassiers, agoons and hussars are listed separately. The anging garrisons are listed as of 1786 with the ain location, since the cavalry was much divided.

Most regiments were stationed in Silesia, then Brandenburg and East and West Prussia, corresponding to the conditions in the extended regions. The hussars had no draft areas, but received volunteers.

BATTLE ACTION OF THE REGIMENTS

MOLLWITZ	April 10, 1741:	C.R. 5, 10 (1 squ.), 11; D.R. 1, 3/4 (8 squ.), 5 (6 squ.); H.R. 2: total: 4500 vs. 9000 Austrians
CHOTUSITZ	May 17, 1742:	C.R. 1, 2, 4, 7, 8, 9, 12; D.R. 3, 5, 7; H.R. 1: total: 7000 vs. 10,000 Austrians
HOHENFRIEDEBERG	June 4, 1745:	C.R. 1, 2, 4, 5, 7, 8, 9, 10, 11, 12, 13 (1 squ.); D.R. 1, 2, 3, 4, 5, 6, 11, 12; H.R. 1 (5 squ.), 2, 4, 5, 6, 8 (5 squ.): total: 19,900 vs. 24,000 Austrians and Saxons
SOOR	September 30, 1745:	C.R. 1, 2, 4, 8, 9, 10, 12, 13 (1 squ.); D.R. 3; H.R. 4 (3 squ.): total: 5900 vs. 16,700 Austrians and Saxons
KESSELSDORF	December 15, 1745:	C.R. 1, 3, 5, 6, 7, 8, 11, 12; D.R. 4, 5, 7, 8, 9, 10; H.R. 6, 7: total: 9000 vs. 7000 Saxons
LOBOSITZ	October 1, 1756:	C.R. 1, 3, 5, 6, 7, 8, 10, 11, 13 (1 squ.); D.R. 3, 4, 8 (5 squ.); H.R. 1 (8 squ.): total: 8700 vs. 7500 Austrians
PRAGUE	May 6, 1757:	C.R. 1, 3, 4, 5, 6, 7, 9, 12, 13; D.R. 1, 2, 3, 4, 11, 12; H.R/ 2, 3, 4, 6: total: 16,500 vs. 13,000 Austrians.
KOLIN	June 18, 1757:	C.R. 1, 2, 3, 6, 7, 8, 11, 12, 13; D.R. 1, 2, 3, 4, 11; H.R. 1 (5 squ.), 2, 3, 4, 6, 8 (5 squ.): total: 14,000 vs. 19,000 Austrians
GROSS-JÄGERSDORF	August 30, 1757:	D.R. 6, 7, 8, 9, 10; H.R. 5, 7, Bosnians: total: 6000 vs. 11,000 Russians
ROSSBACH	November 5, 1757:	C.R. 3, 7, 8, 10, 13; D.R. 3, 4; H.R. 1, 8 (2 squ.): total: 5400 vs. 8500 French and Imperial Army
BRESLAU	November 22, 1757:	C.R. 1, 2, 4, 5, 6, 9, 11, 12; D.R. 1, 2, 5, 11, 12; H.R. 2, 4, 6 (8 squ.), 8 (3 squ.): total: 10,000 vs. 23,225 Austrians
LEUTHEN	December 5, 1757:	C.R.1, 4, 5, 6, 7, 8, 9, 10, 11, 12, 13; D.R. 1, 2, 4, 5, 11, 12; H.R. 1 (5 squ.), 2, 3 (3 squ.), 4, 6 (8 squ.), 8 (5 squ.): total: 11,000 vs. 14,000 Austrians
ZORNDORF	August 25, 1758:	C.R. 2, 5, 8, 10, 11, 13; D.R. 1, 4, 6, 7, 8; H.R. 2, 5, 7: total: 10,500 vs. 6,400 Russians
HOCHKIRCH	October 14, 1758:	C.R. 1, 4, 6, 8, 9, 10, 11, 12, 13; D.R. 1, 2, 4, 5, 11, 12; H.R. 2, 4, 6: total: 13,500 vs. 17,200 Austrians
KAY	July 23, 1759:	C.R. 1, 5, 7, 12; D.R. 6, 8; H.R. 2, 4, 5 (6 squ.), 7 (7 squ.), Bosnians: total: 7800 vs. 8600 Russians
KUNERSDORF	August 12, 1759:	C.R. 1, 2, 3, 5, 7, 12; D.R. 2, 3, 6, 8, 11; H.R. 1 (11 squ.), 2 (6 squ.), 3 (3 squ.), 4, Belling (5 squ.): total: 12,500 vs. 15,000 Russians and Austrians
LIEGNITZ	August 15, 1760:	C.R. 2, 3, 5, 8; D.R. 2; H.R. 2, as well as C.R. 10, 11, 13; D.R. 1, 4, 9, 10; H.R. 3: total: 8600, 3900 in battle
TORGAU	November 3, 1760:	C.R. 1, 2, 3, 4, 5, 8, 10, 11, 12, 13; D.R. 1, 2, 4, 5, 11 (3 squ.), 12; H.R. 1 (2 squ.), Free Hussars, 2, 4 (2 squ.): total: 13,500 vs. 10,700 Austrians

THE KING'S ADDRESS TO HIS GENERALS AND STAFF OFFICERS
on December 4, 1757, the eve of the Battle of Leuthen

Gentlemen, it is known to you that the Prince of orraine has succeeded in capturing Schweidnitz, eating the Duke of Bevern and making himself naster of Breslau, while I have been compelled to estrain the progress of the French and Imperial eople. A part of Silesia, my capital and all my artime needs which were located there have been lost, nd my annoyance would have reached its highest oint if I did not have unlimited trust in your ourage, your steadfastness and love of your country, hich you have proved to me on so many occasions. I ecognize these services to the Fatherland and me om the bottom of my heart.

There is scarcely one of you who has not istinguished himself by some great, honorable eed, and I flatter myself to think that, should the pportunity arise, you would not be lacking in the ravery that the State has a right to require of you.

This point in time is coming; I would feel that I ad done nothing if I left the Austrians in possession f Silesia.

Let me say to you that I intend, against all the rules f the art, that I shall attack Prince Karl's army where find it, though it is three times as strong. It is not a uestion of the numbers of the enemy, nor of the nportance of his chosen position; all this, I hope, ill help to increase the heartiness of my troops and e proper following of my directions.

I must venture this step, or all is lost; we must beat e enemy or let ourselves be buried before his atteries.

So I think—so I shall act.

Make this decision of mine known to all the officers in the army; prepare the common man for the actions that will soon follow, and inform him that I consider myself justified in expecting unqualified obedience of him. If you also bear in mind that you are Prussians, then you will surely not make yourselves unworthy of this advantage; but if there is one or another among you who is afraid to share all dangers with me, he can still obtain his release today without enduring the slightest criticism from me.

I am convinced in advance that none of you would desert me; I am counting completely on your loyal help in the certain victory.

Should I remain and not be able to repay you for your services to me, then the Fatherland must do so.

Go to the camp now and repeat to your regiments what you have heard from me now.

The cavalry regiment that does not immediately charge unceasingly into the enemy when ordered to, I shall demote immediately after the battle and turn into a garrison regiment.

The infantry battalion that begins to come to a stop, no matter what it encounters, will lose its colors and its saber, and I shall have the braid cut off its decorations. Now fare you well, Gentlemen; in a short time we shall have beaten the enemy or we shall never see each other again.

CUIRASSIER REGIMENT 1
Mounted Regiment

Commanders of the Regiment

1665	10/6	Johann Georg, Prince of Anhalt-Dessau, later Field Marshal
1693		Major General Carl Friedrich, Count von Schlippenbach, later General of the Cavalry
1723	7/15	Major General Cuno Ernst von Bredow
1724	7/18	Colonel Wilhelm Dietrich von Buddenbrock, later Field Marshal
1757	4/2	Major General Hans Kaspar von Krockow, died 2/25/1759 of wounds at Hochkirch
1759	2/28	Major General Gustav Albrecht von Schlabrendorff, died 1765
1765-	1768	vacant
1768	12/20	Colonel Friedrich Wilhelm von Roeder
1781	3/19	Major General Levin Friedrich Gideon von Apenburg, later chief of Dragoon Regiment 7
1784	6/13	Major General Philipp Christian von Bohlen, later Lieutenant General
1787	10/14	Major General Georg Dietrich von der Gröben, later Chief of Dept. 5 of the Upper War Collegium
1788	10/30	Colonel Diedrich Goswin von Dolffs, later Lieutenant General
1805	10/17	Colonel Elias Maximilian, Count Henckel von Donnersmarck

When the Bishop of Münster fell on the rear of the Hollanders in the Second English-Dutch Trade War of 1664-1667, the Great Elector, in Cleves, sent 800 newly stationed dragoons from East Prussia and 1000 cavalrymen called from Brandenburg marching westward. In the course of his alliance, he ordered on February 3, 1666, that the cavalry be tripled to 3,500 and the dragoons be doubled to 1500 men. The first of the seven cavalry regiments of 500 men each in six companies was turned over to Prince Johann Georg of Anhalt-Dessau. When Münster gave way and the Jülich-Cleves Compromise was accepted, a strong squadron of 200 men of the regiment remained as a permanent force in Halberstadt territory. On January 24, 1670, the Elector promoted his brother-in-law, Johann Georg, to Field Marshal. The regiment stood beside the Hollanders on the lower Rhine in 1672-1673, filled to 600 men. In 1674 it fought in the Imperial Army at Strassburg and Colmar, while Prince Johann Georg was Governor of Pomerania and the Mark. The Swedish attack on the Uckermark at the end of 1674 called it to the Havelland at the end of June 1675, where it beat them on June 28. In the Pomeranian campaign beginning on October 15, it took Greifenhagen and besieged Stettin under Count Dohna in 1676 and 1677; the city fell on December 27. In 1678 it took Rügen, Stralsund and Greifswald. In 1679, at full strength, it was garrisoned in East Prussia, and in 1686 it provided one company for the supply corps in Hungary, which helped to capture Ofen on July 27. Strengthened to eight companies, it moved against France early in 1689 before Bonn, which fell on October 12. As of March 4, 1691 it provided one company for the founding of von Schöning's Cavalry Regiment 9 for service in Hungary, and fought at full strength in the fall of that year at Steenkerke and Leuze in the West, then on July 29, 1693 at Neerwinden, where it fought in the center and suffered heavy losses. Then for a year it served before the fortresses of Brabant. In 1697 it was reduced to three companies and garrisoned in Anhalt, and later in East Prussia.

In 1704 the companies were enlarged to 75 privates. In the summer of 1706 it went to Flanders under Count Lottum, took Menin, fought at Oudenarde and Wynendael in 1708, beat back the defenders of Ostende and captured Lille and Ghent. In 1709 it acquitted itself bravely despite losses in the Battle of Malplaquet. The regiment was recalled in 1710 and stationed with six companies between the Uckermark and Lebus. It had eighteen officers, 36 junior officers and 480 men, 534 in all. In November of 1715 it took part in the landing on Rügen. In 1718 it received its fourth squadron from the Heiden Regiment, gained its fifth at year's end and now had 32 officers, 60 junior officers and 685 men, plus 30 supernumeraries as of 1727, for a total of 806.

As of 1733 its replacements came from Riesenburg and Marienwerder, as of 1740 from recruitment in Silesia, as of 1743 from Silesian infantry regiments, only as of 1798 from the districts of Oels and Wartenberg, with the cities of Bernstadt, Juliusburg, Festenberg, Goschütz and Hundsfeld. Its garrisons were Bischofswerder, Deutsch-Eylau, Marienwerder, Riesenburg and Rosenberg in West Prussia from 1719 to 1739, as of 1743 Breslau and Schweidnitz, since 1746 only Breslau. Count von Schlippenbach, its longtime chief in two wars, had been brought up since the age of ten with the young King of Sweden and had served as a pikeman for twelve years; in Brandenburg since 1686, he was a clever, brave, perceptive leader. His successor—for 33 years, as of 1724—was Wilhelm Dietrich von Buddenbrock, university-educated, one of the most outstanding cavalry leaders of the eighteenth century, who led the Prussian Cavalry to victory at Chotusitz, Hohenfriedeberg and Soor.

In February and March of 1741 the regiment went to Silesia with the Third Army Corps, did not take part at Mollwitz, but could evaluate all its experiences during the summer. As of late October of that year it secured northern Bohemia until May of 1742 under Hereditary Prince Leopold, going to Chotusitz on May 16. In the first line on the right wing under Buddenbrock, it experienced a ride to death to open the battle. Colonel von Maltzahn, the regimental commander, fell. Two pour-le-merite and specific recognition expressed the King's thanks. As peace came on June 11, it was transferred to Silesia; Buddenbrock trained old cuirassier and dragoon regiments. Attacks were supposed to be set up diagonally, begin at a trot and become a full gallop for the last hundred paces, then immediately form and close for a second attack without any shooting. The new regulation of 1743 brought the gallop to a position of honor and demanded five times as much practice as maneuvers on foot. The Prussian Cavalry was to be the equal of the Austrian!

As of February 15, 1744, preparations for a campaign began again; the regiment joined the King's First Corps and experienced the unfortunate autumn campaign in Bohemia. With the army until the end of May 1745, it took part on the inner right wing of the opening attack at Hohenfriedberg on June 4, which even overrode the grenadier corps. The regiment suffered the greatest losses but captured four flags and four pairs of kettledrums. It received four pour-le-merite. At Soor, again under the 73-year-old Buddenbrock and with von der Goltz's Brigade, it again opened the attack on the right wing against the northeast slope of the Graner Koppe, drove back fifty enemy squadrons and captured the main battery, so that the victory was won despite the difficult terrain. Colonel von Ledebur, the regimental commander, fell. As of December 13 it joined Lehwaldt's Corps and the army of the Old Dessauer in Saxony. At Kesselsdorf it was on the outermost left flank north of Zöllmen, facing the swampy Zschoner Valley. If the cavalry had crossed this hindrance in time, the Saxon Army would have been smashed. Then it spent the next twelve years in its Breslau garrison.

In 1756 the regiment belonged to the Second Corps under Schwerin in Silesia, which made nothing but feints. After enlargement to 969 men as of February 1, 1757, it marched via

1. Kürassier-Regiment
NON SOLI CEDIT
Uffz.
G. Dorn

Jungbunzlau to Prague by May 6. Here it attacked from Unter Poczernitz on the left flank in the middle of the first encounter south of Sterbohol, where the enemy's right flank was beaten with Zieten's help. Only 442 horses strong at Kolin, it attacked the Krzeczhorz Heights west of Brzistwi in the Pennavaire Division without infantry, but was driven back to the Kaiserweg by the Imperial cavalry, with heavy losses. At the end of August it rode to Silesia under the Duke of Bevern; on November 22 it could not avoid defeat at Breslau, where General Pennavaire was fatally wounded. On Decmeber 2 it joined the King at Parchwitz. At Leuthen it took part in the first line of Driesen's final attack on the left wing, from the Radaxdorf Heights to Mühlberg and into the midst of the Austrian front, which was turning to the south. In 1758 it joined the King's Army and experienced the unsuccessful move on Olmütz until early August, remaining in Silesia afterward. At Hochkirch it stood west of the headquarters of Rodewitz and then formed an intercepting line between Drehsa and Parchwitz, white its chief, General von Krockow, was fatally injured on the right wing. In Prince Heinrich's Saxon Corps in 1759, it went to the Pomeranian Corps on June 24 and carried out self-sacrificing attacks against the Palzig Heights at Kay. Its regimental commander, Colonel von Wartenberg, fell. At Kunersdorf three weeks later it lost 107 dead, including twelve officers, in the attack between Kunersdorf and Mühlberg, small attacks to take pressure off the infantry, and the cavalry battle in the enemy's last cavalry attack—all in vain! In 1760 it marched from Dresden to the Warthe, from Breslau to Torgau. Here it led Holstein's Cavalry on the left flank in circling the enemy and making the great attack on the Süptitz Heights, driving back the enemy cavalry wing despite counter-attacks. During 1761 it was involved in the ongoing defense of eastern Saxony. In 1762 it saw the breakthrough on the Mulde, the camp at Pretzschendorf and, in the Bandemer Cuirassier Brigade on October 29, victory at Freiberg in the second attack column between St. Michael and the Spittelwald.

In the War of the Bavarian Succession in 1778 and 1779 it was with the King's Army at Krakau and Posen. In 1806 it capitulated at Pasewalk on October 29, and the Breslau depot surrendered on January 5, 1807. The remaining men joined the First West Prussian Dragoon Regiment.

1. Kürassier-Regiment

CUIRASSIER REGIMENT 2
Mounted Regiment

Commanders of the Regiment

1656	1/28	Colonel Georg Adam von Pfuel, later Major General
1670		Colonel Johann Christoph von Strauss, later Major General, died in 1686 before Ofen.
1672	June	Prince Friedrich, Elector Prince in 1674, ruling Elector in 1688, King in 1701 (and always His Royal Highness the Crown Prince as Colonel)
1730	8/31	Colonel August Wilhelm, Prince of Prussia, since 1744 heir to the throne, later General, died in 1758
1758	12/8	Colonel Heinrich, Prince of Prussia, second son of the above (commissioned only in 1764)
1786	12/20	Colonel Georg Ludwig von Wiersbitzki, later Major General
1778	3/11	Major General Christian Rudolf von Weyherr
1782	9/21	Major General Friedrich August von Saher
1783	3/16	Colonel Karl von Backhoff, later Director of Dept. II of the Upper War Collegium
1789	4/8	Major General Gustav Ludwig von der Marwitz, formerly of C.R. 7
1797	2/25	Colonel Peter Ewald von Malschitzky, later Major General
1802	3/20	Colonel Andreas Dietrich von Schleinitz, later Major General
1805	10/17	Colonel Carl Friedrich Hermann von Beeren

Before the 1656 Swedish-Polish War, the Great Elector empowered Colonel Adam von Pfuel to create a regiment on December 13, 1655. Established with four companies as of January 28, 1656, it was promptly taken along to Poland, then used against the Danes in Near Pomerania in 1659, its strength already doubled. As confirmed on October 6, 1665, it was reorganized as of February 3, 1666, with the addition of three companies of Lieutenant Colonel von Küssow from Pomerania and three of Captain von Arnstedt from Halberstadt. What with the reduction after the Münster confusion, there remained as of June 12 of that year only one squadron permanently stationed in the Neumark. When Pfuel became Commandant of Spandau in 1670, Colonel Christoph von Strauss took command of the regiment, but as early as June of 1672 he turned it over to Prince Friedrich, later the Electoral Prince. Until 1730 when, as King Friedrich II, he took over I.R. 15, it was known as the "Crown Prince's Regiment." In August of 1673 it still had three squadrons in the Mittelmark. In 1674 it fought in Alsace, where it took Wasselsheim Castle late in October. Early in 1675 it hurried back because the Swedes had invaded and drove them back at Fehrbellin on June 28. From July 1676 to the end of 1677 it besieged Stettin. One squadron took part in the landing on Rügen on September 23, 1678; then it wintered in the Newmark. In 1683 the regiment was expanded to 1500 men in eight companies. In 1686 one company went along to the taking of Ofen, where Major General von Strauss was mortally wounded. In 1689 it was on the Rhine, taking part in the capture of Bonn on October 12, before returning home. In 1690 it went back to Flanders. In 1691 it was strengthened by three companies, but contributed two companies to the formation of C.R. 6. Next it saw service with six companies in Flanders, at Leuze and Namur, and Luxembourg. With eight companies in 1692-1693, it was in Flanders under its commander, Colonel du Rosey, serving at Maastricht, Düren, Tirlemont and finally Heilbronn. In 1694, expanded to nine companies, it captured Huy, in 1695 Namur, and in 1697 it was at Deinze, not far from Ghent, always wintering in the Magdeburg or Halberstadt area. Then it filled its ranks with three companies from the regiment on horseback of von Thümen (formerly Kannenberg, Margrave Ludwig and Lüttwitz).

In the first years of the Eighteenth Century the regiment was stationed in East Prussia. In March of 1705 it joined Arnim's Corps and went to the Mosel, then to Lauterburg to take part in the capture of Hagenau, and then to march back; in 1706 it joined Lottum's Corps and fought at Menin and Ath, with winter quarters between the Maas and Rhine. The year of 1707 was uneventful.

Under Prince Eugene it fought at Oudenarde on July 11, 1708, on the right flank in the first battle line. With Lottum it advanced on Lille, which fell on October 22. Shortly after that one squadron was taken prisoner at Hondschoote. At year's end Ghent was taken; the winter was spent on the Rhine. At Malplaquet on September 11, it took part successfully in the changing cavalry battle after the capture of the battlements. The Prussian Cavalry lost 523 soldiers. In 1710 it took part in the capture of Douay, Bethune and Aire, while its chief, Prince Friedrich Wilhelm I, increasingly made his influence felt on the army. At the end of 1712 the entire campaign ended in winter quarters in the Archdiocese of Cologne. The new King began his reign in 1713 with the words: "I am the Finance Minister and Field Marshal of the King of Prussia; the King of Prussia will confirm that!" The regiment still consisted of three squadrons, of six companies. In 1715 it had to serve in the Pomeranian campaign, where it was involved in the taking of Stralsund. On August 22, 1718 it received two companies from the disbanded regiment of von Wartensleben, founded in 1686, and created two more, so that with five squadrons it now numbered 31 officers, 60 non-commissioned officers and 715 men, a total of 806 soldiers. In 1730 Prince August Wilhelm (1722-1758), the oldest brother of Friedrich II and then heir to the throne, became the regimental commander. After Kolin, the King deposed him as commander of an army corps. From 1733 to 1806 its replacements came from parts of the districts of Ruppin and Havelberg with the cities of Kyritz, Wittstock, Perleberg, Pritzwalk, Wusterhausen, Gransee, Kremmen and Putlitz. Though its garrisons had been in Westphalia until 1720, they were moved into its draft area as of 1721 so that garrison and canton were one and the same until 1806.

This regiment also went to Silesia with the Third Corps in 1741 and had time "to practice real service in the campaign", as the King wanted. At Chotusitz on May 17, 1742, it showed its ability on the left wing in the great attack eastward past the village into the enemy's rear. With C.R. 7 and 12 it rode down two infantry regiments, captured a flag and fought its way through all three Austrian battle lines, disposing of a hussar regiment in the process. Twenty-four of its officers fell; the eight survivors, including Colonel von Katzler, received the Pour-le-merite. In the King's First Corps in 1744 it experienced the unfortunate advance into northern Bohemia; the army had fully recovered by April of 1745. After the King's waiting-it-out tactics along the mountains, the Battle of Hohenfriedeberg took place on June 4. In the middle of the right flank it attacked at Pilgramshain. Within sight of the enemy a lieutenant and twenty men dismounted and put up a board fence, the cover for the Saxon infantry. The regiment set out over the ditches and threw the infantry regiment in a heap, captured a pair of kettledrums, and lost two officers and 42 men. At Soor it joined Cuirassier Regiments 1 and 10 to break through the enemy battle line and take the Graner Koppe with 22 guns.

On August 29, 1756 it marched with the King's First Corps, under Bevern, until east of Dresden and then up the Elbe. At

2. Kürassier – Regiment
PRO GLORIA ET PATRIA
Uffz

Lobositz three of its squadrons rode at the head of the second attack of the right wing from the Homolka-Berg to the southern edge of Lobositz, losing ten officers and 127 men. After the squadrons were strengthened, it fought under Bevern at Reichenberg on April 21, breaking through but also enclosing the small side of Prague under Keith. At Kolin it led the victory ride of the Krosigk Brigade north of Krzeczhorz, overriding three regiments. The breakthrough was attained, the infantry gained a respite. When the heights were not gained and the advantage was not utilized, the enemy cavalry reserve drove it to flight. Then it moved to Silesia; at Breslau on November 22 it could not avoid defeat. In 1758 it covered the siege of Schweidnitz and left on April 20 to join the Saxon Corps to reach Frankfurt/Oder early in August. At Zorndorff it survived the flank attack of the Russian cavalry under Demiku and led the successful counter-attack under General von Schorlemmer on Dohna's inner right wing along with Seydlitz, winning the victory. Then the King took it in the direction of Küstrin. At the end of the year Prince Heinrich of Prussia, second son of Prince August Wilhelm, took over as regimental commander. After the 1759 advance of the Saxon Corps against the magazine in Franconia, it joined the King at Sagan on July 29, crossed the Oder on August 11 and was thrown to the left in the first line on the right wing in the Battle of Kunersdorf, to lose 206 men in luckless small attacks in the narrow space between the village and the Mühlberg. It all ended in flight. In 1760 it fought first at Dresden, then at Liegnitz, where it first met the enemy behind the Anhalt-Bernburg Infantry Regiment and immediately took a battery, then swung to the right toward the Katzbach. Lieutenant Colonel von Wiersbitzki received the Pour-le-merite. On October 7 it began a two-week march from Schweidnitz to Wittenberg. With Zieten's Corps at Torgau, it first protected the flank from Lacy's Corps, then led the way up the Süptitz Heights. In 1761 it carried out raids in Thuringia: Langensalza and Rudolstadt, then fought in Poland. Strengthened by a thousand men in 1762, it fought at Burkersdorf, Leutmannsdorf and Reichenbach. After that it fought in Saxony with Hussar Regiment 2 at Spechthausen, where 600 prisoners and four guns fell into its hands.

In 1763 the regiment consisted of 641 Prussians, 43 Saxons and 269 other 'foreigners', 953 soldiers in all. It came through the War of the Bavarian Succession in the army of Prince Heinrich. Until 1767 the regiment had belonged to the House of Hohenzollern. In the spring of 1806 it moved against the Swedes in Lauenburg and Mecklenburg. In the war against France the Third Squadron capitulated in Erfurt on October 16, the other four with Blöcher's Corps at Ratekau on November 7; the depot in East Prussia then joined the Brandenburg Cuirassier Regiment, later Cuirassier Regiment 6.

2. Kürassier Regiment

CUIRASSIER REGIMENT 3
Mounted Bodyguard Regiment

Commanders of the Regiment

1672		Colonel Nikolaus von Below
1673		Major General Ulrich, Count von Promnitz
1679		Colonel Hans von Sydow
1679		Colonel Joachim Balthasar von Dewitz, later Lieutenant General and Governor of Kolberg
1695		Lieutenant General von Wangenheim
1709		Major General Wolf Christoph von Hackeborn
1719	May	Major General Gottfried Albrecht von Bredow, later Governor in Peitz
1726	1/20	Major General Friedrich Wilhelm von Dewitz, later Lieutenant General
1736	11/25	Colonel Adam Friedrich von Wreech, later Lieutenant General
1746	9/25	Major General Nikolaus Andreas von Katzler, later Commander of the Gensdarmes
1747	9/20	Major General Hans Friedrich von Katte, later Lieutenant General
1758	1/5	Major General Robert Scipio, Baron von Lentulus, later Lieutenant General
1778	12/22	Colonel Johann Rudolf von Merian, later Major General
1782	9/21	Major General Ernst Christian von Kospoth, later Lieutenant General
1794	12/29	Major General Heinrich Leopold, Count von der Goltz
1797	2/25	Colonel August Friedrich von der Drössel, later Major General
1799	6/15	Major General Ulrich Carl von Froreich
1801	3/24	Colonel Hermann von Kölichen, later Major General
1805	10/17	Colonel Friedrich August Carl Leopold, Count von Schwerin, later Major General

Since 1656 the mounted regiment of Colonel Alexander von Spaen, organized for the war against Poland and consisting of four companies, had been known as the "Bodyguard Regiment on Horseback". Despite its changing composition, it had been authorized for General Spaen in 1666. In 1672 Colonel Nikolaus von Below formed a new regiment which was elevated to the "Bodyguard" status after its formation and soon enlarged to six companies. The Spaen Regiment was divided in 1718 to form Heiden's Regiment. But no special activity went along with the honorary title, other than a direct connection to the Prince and "Commander-in-Chief" instead of a regimental chief of its own. Obviously it was not often used in battle, and least of all farmed out to foreign service. In 1674 it joined the unsuccessful autumn campaign in upper Alsace, and in 1675, after wintering in Franconia and Thuringia, it joined the Elector's memorable march with his mixed mounted army from Magdeburg to Fehrbellin in five days to take part in the victory over the Swedes. Then Stettin was cut off from the mouth of the Oder and extensively surrounded, until it capitulated on December 27, 1677. Before that it took Anklam on August 29, 1676. In 1678, under Homburg, it went to East Prussia, but was back in Near Pomerania in mid-September and sent one company to land in Rügen. In the winter of 1678-1679 it was in the Vogtland, until Brandenburg had to give up all its gains in Pomerania by the end of 1679, according to the peace treaty. In 1686 it sent one company to Hungary to help storm Ofen. In 1689 it was with the army of the Saxon Electorate for a time, and gained two more companies in the spring of 1689, when it returned. Three squadrons of three companies each were planned for by 1691. In 1689, after the loss of Kaiserswerth, it took part in the conquest of Bonn on October 12. In 1691 it contributed one company to the founding of Cuirassier Regiment 9, and in mid-May it was near Brussels and Huy for the advance to Luxembourg. In 1693 it secured the area between the Maas and Ourthe. In the Battle of Neerwinden on July 29 its closing attack on the French could not stave off defeat. Then came the retreat to Louvain. In mid-July of 1694 it gathered at Louvain again, and as of June 1695 it stood before Namur, which was taken on September 2. In the peacetime budget of 1697 it was reduced from six companies, and in 1701 it gave another thirty men to Cuirassier Company 9.

From 1703 to the end of 1705 it secured threatened East Prussia. In 1706 it marched west again under Lottum, helped to take Menin and, in mid-September, Ath in Flanders. The year of 1707 passed with only insignificant action. In the Battle of Oudenarde in 1808, it successfully attacked the French on the plateau north of Eyne, took Menin again and captured the Lys crossings. From August 22 to October 22 it besieged Lille, and at year's end it captured Ghent. In 1709 it successfully besieged Tournay from the end of June to July 29. On September 11 it fought in the great cavalry battle of Malplaquet, winning a victory in the end despite losing 523 men. After capturing Mons on October 20 it wintered in Geldern. In 1710 Prince Leopold of Anhalt-Dessau took command of the corps and captured Douzy between May 4 and June 26, Bethune on August 29, and Aire on November 9. In 1712 it fought at Landrecy from July 16 to 31, after protecting Le Quesnoy on July 2 during the siege. When peace was made in August of 1713, it had three squadrons with six companies. King Friedrich Wilhelm I proclaimed a draft law on May 9, 1714: "The young man, according to birth and God's order and command, is liable and obligated to serve with life and property!" During the 1715 siege of Stralsund, which the Swedes defended, two of its squadrons saw action in the conquest of Rügen on November 18. Stralsund did not fall until December 23 of that year. In 1718 it was reorganized with five squadrons, including one from the Wartensleben Regiment. Replacements for the regiment came from the districts of Aschersleben, Oschersleben and Holzkreis with the cities of Schönebeck/Elbe, Frohse, Salze, Mansfeld, Hammersleben and Gerbstedt from 1733 through 1806. Since 1714 its garrisons had been in the areas of Salzwedel, Tangermünde, Heiligenfelde and, as of 1720, also Arendsee, Gardelegen and Kalbe/Altmark, and from 1732 to 1806 in its canton as well as the towns of Alsleben, Seehausen, Wanzleben and Heimersleben, replaced by Egeln from 1746 on.

In April of 1741 the regiment joined the Old Dessauer's Observation Corps in camp at Göttin, south of Brandenburg/Havel, ready to take on Saxony and Hannover. In the event that Russia intervened, Saxony was to be defeated before Hannover could enter the fray. According to Mollwitz's experiences, Prince Leopold also told the cavalry to "attack at a strong trot, even a short gallop, and then because your horses are bigger than the others, you absolutely must smash everything down". In attacking at a gallop, order and control of the horses could not be neglected. The mounted regiments, now often known as "cuirassiers", were used in the first battle line from now on as a matter of principle, the dragoons in the second, and the newly-established hussars placed behind them and on the wings in the battle line. Transferred to Grüningen as of September 12, the corps was disbanded at the end of October. There was no threat of danger from Saxony in 1742. On September 25, 1744 the Dessauer was given command of the central provinces because of Saxony's hostile position, and at the end of April he gathered a weak corps at Magdeburg, including the regiment. It moved forward slowly, reaching Dieskau near Halle by the end of August, and was twice strengthened with troops from Silesia. The corps was on the alert since mid-November. When Lehwaldt's Corps arrived on December 13, the Prince advanced to Kesselsdorf, where it was used on the extreme right flank in the first encounter, flanking Kesselsdorf, the key to the enemy position. It lost three officers, 28 men and 84 horses.

The regiment belonged to the King's Army that marched into Saxony on August 29, 1756 and moved upstream west of the river on September 6. On October 1 it rode in the first main attack of

3. Kürassier-Regiment
NON SOLI CEDIT
Uffz.
G. Dorn

the first line on the left wing against Lobositz and the Modl Brook, sweeping the Austrians off the field anew. In mid-April 1757 it went to Prague with the King. On May 6 at Hrdlorzez, after the infantry's success it broke into a gallop on the right flank despite the rough terrain and made victory certain. Six weeks later at Kolin, in Oennavaire's Division, it plunged into the gap between the Krzeczhorz Heights and Eichbusch, west of Brzistwi, moving uphill to reach the top, only to be thrown back by the counterattack. Five weeks later it marched via Bautzen, Naumburg and Gotha to Leipzig. At Rossbach on November 5, under Seydlitz,it showed its ability particularly well in the first battle line on the left in both attacks on the enemy cavalry and then against the infantry, "and earned the King's favor". Only part of the Prussian infantry entered the battle. In 1758 it took part in the unsuccessful advance to capture Olmütz, then joined the Saxon Corps, which defended itself capably at Gross Sedlitz at the end of August, Dresden in mid-September and Eilenburg in November. In 1758 General Lentulus, a friend of the King, became its chief and remained so for twenty years. When Prince Heinrich destroyed the magazines at Saaz and Budin in the Bohemian border area in April of 1759, two squadrons joined the hussars, dragoons and grenadiers under Lieutenant Colonel von Belling to shatter three enemy battalions and capture 800 men, one general, three cannon, eight flags and three standards. On July 30 it joined the King at Sagan, crossed the Oder on August 11 and went to Kunersdorf where, on the right wing in Finck's Corps, it advanced over the Trettin Heights from the north. In the great cavalry battle it attacked without success over unfavorable terrain, with many losses. Two squadrons got into the swampy Elsbusch west of the Mühlberg and lost many prisoners and a standard. Seven officers and 98 men fell. From July 10 to 22, 1760 it fought in the King's Army before Dresden, the recapture of which failed, and on August 15 at Liegnitz it was on the left wing to meet the attack of three enemy cavalry regiments, 1, 24 and 35, of the avant-garde from Bienowitz, capturing seven flags and five cannon. On October 7 it left for the west; at Torgau on November 3 it was part of Holstein's cavalry on the left wing, attacking the Süptitz Heights from the north in the afternoon but having to retreat under flank fire and a counterattack. In 1761-1762 it was in Saxony again and advanced to Langelsalza under Löllhöfel and to Schwarzburg—Rudolstadt under Sydow.

After the 1778-1779 campaign it advanced into the Saarland again in 1793 against the French. In 1806 most of the regiment capitulated at Prenzlau on October 28, parts at Anklam and Wahren on November 1. The rest later joined the Brandenburg Cuirassier Regiment.

3. Kürassier-Regiment

CUIRASSIER REGIMENT 4
Mounted Regiment

Commanders of the Regiment

1674	mid-	Colonel Joachim Ernst von Grumbkow, later War Commissioner General
1682	10/4	Colonel Dietrich, Count zu Dohna, died 7/27/1686
1686		Colonel Joachim Friedrich von Wreech, later Lieutenant General
1714	Feb.	Brigadier Peter von Blanckensee
1733	5/8	Colonel Friedrich Leopold von Gessler, later Count and Field Marshal
1758	1/5	Major General Johann Ernst von Schmettau
1764	9/1	Major General Hans Georg Woldeck von Arneburg
1769	6/11	Colonel Georg Christoph von Arnim, later Lieutenant General
1785	9/23	Major General Carl Friedrich, Baron von Mengden, formerly of the Garde du Corps, later Lieutenant General
1796	8/22	Major General Carl Friedrich Ernst Count zu Waldburg, Hereditary Steward
1800	6/11	Major General Ernst Philipp von Wagenfeld

In September and October of 1672 Colonel Joachim Ernst von Grumbkow organized two companies of 100 men each for ordnance and guard duty at the Electoral court; they became known sarcastically as the "Court and Kitchen Dragoons". But in mid-1674 von Grumbkow organized a company of dragoon guards for the field army; it was doubled at the end of the year, and early in 1685 it was brought up to regimental strength with four, then six companies. A direct connection with his first founding is not provable, but can be imagined. After brief use in Alsace at the end of 1674, it went back to Brandenburg immediately. On the eve of the Battle of Fehrbellin it came from Berlin to join the army, taking the village the next day and capturing four flags and five cannon. In the war in Pomerania it first passed by Stralsund and Wolgast to its winter quarters east of the Oder; in early July of 1676 it moved to Teterow, then via Anklam to Stettin. On July 1, 1677 it was strengthened by two companies recruited in East Prussia, and was now called "Bodyguard Dragoons". At the end of 1677 Stettin surrendered. In the landing on Rügen on September 23, 1678 it took 200 prisoners under Colonel von Grumbkow, who had taken the position of War Commissioner General for army supply as of that May. After Stralsund fell on October 25, it took up winter quarters in Halberstadt with eight companies, numbering 512 privates. On October 2, 1682 Grumbkow turned the regiment over to Colonel Dietrich, Count zu Duhna, who led it—half of its men being from the Derfflinger Dragoons—into battle at Ofen in 1686 and fell in the storming on July 27. By September 2 it had taken two horsetails and nine flags, but lost 215 of its 324 men. At the beginning of 1689 eight of its companies gathered on the lower Rhine. After fighting at Urdingen it rescued the Elector from a perilous situation on July 21. Bonn fell on October 12. In the winter of 1689-1690 it was back. The reason for its division into so many supply corps was the State's financial shortage. In 1690 it was in Brabant, at Brussels. In 1691 it protected Liege, experienced the defeat of Leuze and the advance into Luxembourg. In 1692 it led the fight at Mehaigne that avoided a battle, was stationed at Huy and relieved Charleroi. In 1693 it was part of the occupation of Liege and took a beating at an advance post near Tongern on July 15. Huy was lost on July 23. In 1694 it held Liege and fought in the retaking of Huy until September 27. From there two companies went to protect Cleves, under Seckendorff. In 1695 four squadrons took part in the capture of Namur from June 19 to September 2. In 1696 it returned to the Mark Brandenburg. In peacetime the regiment maintained six companies.

As of May 1701 it belonged to Heiden's supply corps, with its companies now numbering 60 men each, but only arrived in Geldern as of April 21, 1703; from there it marched south under Prince Leopold of Anhalt-Dessau to join the Imperial Army, "an elite and fresh troop". In the defeat of Höchstädt on September 20, it defended the crossing of the brook at Ober-Glauheim on foot, losing six officers, 128 men and 200 horses. It also smashed six enemy squadrons and captured two standards. It justified the army commander's confidence in it: "The Bluecoats won't leave me in the lurch!" Nevertheless it lost its status as "Bodyguard Regiment" on February 27, 1704. In that year it went from the Rhine to the Danube in Anhalt-Dessau's Corps. At Höchstädt on August 13 it fought in the main attack of a changing battle southeast of Unter-Glauheim, winning a historic victory. In the process it lost almost all its officers, including its commander, Colonel Ludwig, Count von Blumenthal, 300 men and two standards, but captured one flag and two standards. In 1705 four squadrons joined Arnim's Corps and reached the Mosel by mid-July, fought at Lauterburg and helped to take Hagenau on October 6. In 1706 it besieged Menin from July 23 to August 23, then Ath, and returned to the Rhine for the winter. Nothing eventful happened in 1707. In 1708 Prince Eugene put Lottum's Corps at the head of the army when he opened the Battle of Oudenarde on July 11; in the attack over the highlands at Müllem it lost five officers and 120 men, again the Prussians' heaviest losses. A standard and a pair of silver carabinier kettledrums were modest trophies. Then Lille and Ghent were taken. In 1709 it was present at the taking of Tournay, and on September 11 it survived the bloody fighting for victory at Malplaquet. In 1710 the conquest of the fortresses of Douay, Bethune and Aire cost it many lives. In 1711 it was with Anhalt-Dessau after the skirmish at Estrum, spending six weeks before Bouchain until September 13. In 1712 it stood before Landrecies as of July 16, until the war finally came to an end. In 1715 two squadrons helped to enclose Wismar from mid-June on, and one squadron landed at Gross-Stresau, east of Putbus on Rügen, on November 15. Then it was part of the corps that besieged Stralsund, which fell on December 23, after which the regiment marched off to winter quarters. Early in 1718 it was called a "Regiment on Horseback", and that August it was strengthened by two companies. Until 1742 its replacements came from Mohrungen, and also from the district of Elbing from 1739 on, through recruitment in Silesia as of 1743, from the districts of Neustadt and Oberglogau in Upper Silesia as of 1748, and from the districts of Radom and Pilica and parts of Warsaw from 1796 on. Its garrisons were Lyck, Johannisburg and Lötzen in 1714, Köslin and Stolp in 1716, Mohrungen, Preussisch Holland, Ortelsburg and Lötzen in 1719, also Neidenburg as of 1737, then Rügenwalde, Belgard and Köslin in 1740, Landsberg/Warthe in 1741, Liegnitz in 1743, Ratibor, Leobschütz, Pless and Neisse in 1746, Neustadt, Zülz, Ziegenhals and Oberglogau in 1748, plus Krappitz in 1764, and Warsaw as of 1797.

In 1741 it again belonged to the Third Corps, which moved into Silesia only that spring. The King's experience at Mollwitz taught him to keep the mass of the cavalry on one wing in order to smash the enemy cavalry and then to tear up the unprotected flank of the infantry. On September 29 the regiment led the first victorious fight at Hermsdorf-Kalteck. Then it surrounded Glatz

4. Kürassier-Regiment
NON SOLI CEDIT
Standartenträger
G. Dorn

under Hereditary Prince Leopold, and at the beginning of 1742 it was on the Elbe line. On May 13 it reached the King's camp at Chrudim. On the right flank in the main attack at Chotusitz, it moved against the enemy's left wing of cavalry until it met the second battle line, which the Prussian left wing then surrounded. Here it lost seven officers, 39 men and 47 horses, but received a Pour-le-merite. In 1744 it fought at Prague and at Jaromirz on the way home. Early in 1745 it repulsed a dragoon regiment at Jägerndorf. On May 22 it proved itself at Neustadt under Margrave Carl, earning three Pour-le-merite. At Hohenfriedeberg under Nassau, it won the left-flank cavalry battle before Thomaswaldau and lost two officers and forty men. On the extreme left wing at Soor it led the final attack southeast of Burkersdorf. In 1756 it went to Silesia with Schwerin's Corps and took part in the six-week attack from the County of Glatz, then marched to Prague via Jungbunzlau, arriving on May 6. In the first line of the left wing it opened a pond dam south of Sterbohol despite heavy artillery fire, attacked three times under Prince Schönaich, and gained time for the other regiments to regroup so Zieten could lead the attack that won the victory. It lost eleven officers, 101 men and 83 horses. Then it guarded the besiegers of the city and protected the retreat at Kolin. In Silesia with Bevern it came through the cannonade of Barschdorff, guarded Schweidnitz in vain, and could not avoid defeat at Breslau on November 22. At Leuthen it took part in the second line of the outer left cavalry wing, including Driesen's final flank attack from Radaxdorf, which ended the battle. Six officers and 51 men fell, and an enemy standard was captured by Major von Oginski.

In 1758 it became a "Cuirassier Regiment" and was back in the King's army before Olmütz. At the end of June it helped to protect the supply train from Troppau that was attacked at Domstadtl. Three officers and thirty men fell. At Hochkirch it held its ground on the left flank and formed the infantry's intercepting position. Then it went to relieve the fortress of Neisse, followed by the march into Dresden. In 1759 it was encamped with the King at Schmottseiffen. When he went to Kunersdorf, it joined Fouqu's Corps to fight at Troppau, then at Sorau under Prince Heinrich on September 2. At Hoyerswerda on September 25 it captured General Vehla, 28 officers and 1785 men. After that it marched to Saxony. At Kossdorf on February 20, 1760 it was attacked by enemy forces and drove them off, losing four officers and 67 men while accounting for 120 enemy losses. After the encounter at Bublitz it went into the Battle of Torgau in Holstein's column, attacking the Süptitz Heights after the second infantry attack. In 1762 it fought in the breakthrough at Döbeln on May 12, suffered the loss of ten officers and 317 men in its advance post at Chemnitz, and took revenge at Freiberg: with the cry of "Chemnitz!" it smashed two regiments at the Spittelwald, captured ten guns and eight flags, and pursued the enemy to behind Freiberg, losing only three officers and 21 men. Seven Pour-le-merite honored the regiment's most glorious day.

In the reserve corps in East Prussia in 1806-1807, it later joined Cuirassier Regiment 1.

4. Kürassier-Regiment

CUIRASSIER REGIMENT 5
Regiment on Horseback

Commanders of the Regiment

1683	May	Major General Heinrich, Baron de Briquemault, Sieur de St. Loup, later Lieutenant General, died 8/16/1692
1692		Major General Johann Sigismund, Baron von Heiden
1692	10/25	Major General Philipp Wilhelm, Margrave of Brandenburg, died 12/19/1711 as Grand Master of the Artillery and Lieutenant General
1712	3/31	Colonel Friedrich, Margrave of Brandenburg-Schwedt, son of the above, died 5/3/1771 as Lieutenant General
1771	5/4	Colonel Friedrich Wilhelm Lölhöffel von Löwensprung, Commander-in-Chief since 2/9/1763, later Major General
1780	2/16	Colonel Maximilian von Mauschwitz, later Major General
1782	3/16	Colonel Ludwig Alexander, Prince of Württemberg, later Field Marshal
1800	10/2	Colonel Abraham von Bailliodz, later Major General

In view of the French advances into Alsace, the Great Elector had a mounted regiment established by recruitment in May of 1683; it had six companies, 384 privates, under the command of Major General Henri, Baron de Briquemault, Sieur de St. Loup, and including what remained of Captain von Isselstein's Company, which was the remnant of the Lütke Mounted Regiment that was reduced in 1680. In 1686 it provided 144 men and four company cadre troops for Hungary. At the same time, using mostly transferred German soldiers, it formed ten companies, each with 35 privates, to provide officers' positions for as great a number of French officers as possible, who were to be used at the equal of their French ranks as they were refugees. Here too French spirit blended with Prussian discipline. In 1688 each company had 40 privates. Six companies, 300 privates then went into the service of The Netherlands, while the remaining four companies with 88 privates became the cadre of the new mounted regiment under Major General du Hamel (later C.R. 9), whose regiment had ceased to exist in 1679. The new regiment was intended for use in the 1689 campaign. Normalized that January to five companies with 625 privates, it survived the defeat of Fleuris on July 1, 1690, in Netherlands service. In 1691 it gave one company to C.R. 9. On August 3, 1692 it saw no action in the Battle of Steenkerken, as the cavalry was not used. In the defeat of Neerwinden on July 29, 1693 "it held off more powerful forces bravely". Major General von Heiden was badly wounded. In 1694 three squadrons went from Maaseyck via Louvain to before Huy, which was retaken on September 27. The year of 1695 was dominated by the recapture of Namur from July 2 to September 2. In 1697 the Heyne regiment was absorbed; it had been founded as Derfflinger's in 1666. Thus it retained six companies in peacetime.

In July and August of 1700 it passed for a time to Brandt's Reconnaissance Corps near Lenzen on the Elbe. In 1701 it provided 30 men for Heiden's mounted regiment, which was thereby filled. Since 1692 it was commanded by members of the House of Brandenburg-Schwedt, Margrave Philipp (1669-1711) and Margrave Friedrich (1700-1771), the "Mad Schwedter", half-brother of the first King, cousin of the second and uncle of the third and thus at least moderately related to the House of Hohenzollern. Margrave Friedrich held this position of honor for sixty years!

Under Heiden in 1702, it went along to Kaiserswerth, which surrendered on June 15, then across the Rhine to Venloo, which fell on September 22, then to Roermond, which capitulated on October 7, and finally to Liege, which was taken on October 29. Only Rheinsberg held out until February 7, 1703. The regiment now had three squadrons of 110 privates each. On July 1 it marched to southern Germany under Prince Leopold of Anhalt-Dessau, where it "protected the withdrawal with firmness" in the defeat at Höchs'tdt on September 20. In 1704 the companies of the regiment were enlarged to 75 men. After the strengthening of the Prussian Corps on the Danube, the Allies triumphed together in the Battle of Höchstädt on August 13. In a very changing cavalry battle between Unterglauheim and Blindheim, which was the first decisive event of the battle, it captured a standard and many prisoners. In 1705, after leaving winter quarters at Cham, it served with Marlborough on the Mosel in Arnim's Corps from mid-June on, then was in camp at Lauterburg as of July 25, experienced the breakthrough south-east of Tienen, went to Darmstadt and took part in the taking of Hagenau on October 6. In 1706-1707 it secured East Prussia under Arnimand the Duke of Holstein-Beck. It returned to the west when Prince Eugene took command in 1708. In the victorious battle of Oudenarde on July 11 it rode in the great attack between Müllem and Herlchem under Natzmer. Then it fought at Menin, took the Lys crossings at Comines and Warneton, besieged Lille from August 22 to October 22, and then occupied Ghent. In 1709 it was involved in the taking of Tournay and the encircling of Mons before being used in the victorious cavalry battle at Malplaquet on September 11. It took Douayin 1710—the attack on Arras did not succeed—, then Bethune, and St. Venant on September 30. After the siege of Bouchain in 1711 it moved into winter quarters in the Cologne area; the campaign was practically over for the regiment.

In January of 1713 the companies were again cut to 65 privates apiece. When Friedrich Wilhelm I came to the throne on February 25, he told his ministers: "My father took pleasure in the most extreme magnificence. Allow me to have my pleasure too, which consists mainly of a lot of good troops". He had experienced the defenselessness of the country. In the Pomeranian campaign of 1715 it took part with three squadrons, standing before Wismar under von der Albe in June, then taking part in the siege of Stralsund from mid-July to December 23. In 1718 it received two companies from the disbanded mounted regiment of von Wartensleben, led by Prince Heinrich of Saxony as of 1686, as well as two newly constituted ones, so that it now had five squadrons. Its replacements came from Frankfurt/Oder and vicinity until 1742, then until 1806 from the districts of Schivelbein and parts of Dramburg, and the cities of Schwedt, Angermünde, Falkenberg and Neustadt-Eberswalde, plus Cammin as of 1798. Its garrisons were Kalkar, Goch, Emmerich, Xanten and Mühlheim/Ruhr until 1724, then its canton until 1739, and as of 1743 Belgard, Dramburg, Neustettin, Polzin, Labes, Körlin and Rummelsburg, as of 1788 Treptow/Rega, Wollin, Dramburg and Greifenberg.

On December 16, 1740 it marched into Silesia with the King's First Corps. The new King gave orders about the first invalids on February 11, 1741: "Whenever I see the regiment on review, I want to see them too, and they are to march between the grenadier company and the bodyguard company. I do not doubt that eneryone will be cheered by serving me so much more loyally and happily at all times". At Mollwitz it fought on the inner left wing and lost six officers and 107 men when an

5. Kürassier-Regiment
PRO GLORIA ET PATRIA
G. Dorn

Austrian cavalry attack broke through it. Two weeks later the King wrote: "The cavalry is not worth being taken by the devil". In 1742 it was used in Upper Silesia after serving at Neisse and Ottmachau in 1741. On May 20 it lost 200 prisoners and two standards in an attack at Kranowitz, south of Ratibor. In peacetime the regulations of 1743 allowed only 24 furloughed men per squadron; the regiments had to be ready to march not only in the eight-week drill time, but all year. In 1744 it went with the First Corps to take Prague within four weeks, followed by a withdrawal to Silesia with much privation but few losses. At Hohenfriedeberg it was in the second encounter at the end of the left flank near Thomaswaldau, facing a swampy terrain. But Nassau was able to lead his closed attack of 25 squadrons over Teichau, opening the way for the Bayreuth Dragoons. It lost three officers and 39 men. The King recognized: "—and our cavalry has made us quite respected by the enemy". On December 13, 1745 it went with Lehwaldt's Corps to join the Dessauer at Meissen. In the Battle of Kesselsdorf it belonged to the outer left wing, which saw little action because of the Zschoner Brook north of Zöllmen. In 1756 it marched with the King to Lobositz and lost its commander, Major General von Lüderitz, ten officers and 128 men in the second attack of the first encounter. In 1757 it went to Prague with the King via the Eger and Moldau. Despite heavy losses, it took part in the final attack on the right wing. Then it protected the besiegers of Prague and went to Silesia under Bevern. Under Kyau at Breslau on November 22 it flung itself vainly against superior enemy forces. At Leuthen it fought under Zieten in the first line of the right wing; not far from Gohlau it met Nadasdy's Corps, whose infantry was beaten after a fluctuating cavalry battle. After Schweidnitz had capitulated on April 16, 1758 it joined the Saxon Corps. Taken to Dohna's Corps on the Oder early in August, it defended itself under Schorlemmer in the first line of the right wing at Zorndorf, fighting off Demiku's dangerous attacks, driving two dragoon regiments to flight and cutting down a battalion with sabers. Its commander, Major General Hans Sigismund von Zieten, two officers and 56 men fell. After advances from Saxony to Thuringia in March of 1759, to Bohemia in April and to Bamberg and Bayreuth in May, it went under Hülsen to share in Wedell's defeat at Kay. On the right wing at Kunersdorf it lost 170 soldiers attacking the battlements on the Great Spitzberg. It was thrown into battle at Korbitz, near Meissen, but remained steadfast. In 1760 it came through the bombardment of Dresden and fought on the second line of the left wing at Liegnitz, where it drove back the enemy cavalry and was given a Pour-le-merite for capturing nine guns and ten flags. At Torgau it attacked the Süptitz Heights from the north in the second line of Holstein's Cavalry, cut I.R. 26 and 28 down and captured five guns and three flags, for which it was rewarded with the King's praise and two Pour-le-merite. In 1761 it defended the Mulde line with the Saxon Corps. After the defeat at Brand on October 15, itwent into battle at Freiberg in the second column, under Bandemer, against the northern part of the Spittelwald, pursuing the enemy to the Mulde in the end.

In 1806 it suffered extraordinary losses at Auerstedt. The remaining men went to East Prussia, the depot force to Kolberg and one reserve squadron to Danzig. In 1914 it was the First Brandenburg Dragoon Regiment No. 2.

5. Kürassier-Regiment
G. Dorn

CUIRASSIER REGIMENT 6
Mounted Regiment

Commanders of the Regiment

1689	2/19	Major General Franz du Hamel
1702	3/1	Major General Charles, Count d'Ostange
1704	11/30	Colonel Benjamin Hieronimus Courold du Portail, later Lieutenant General
1715	11/30	Lieutenant General Wilhelm Gustav, Hereditary Prince of Anhalt-Dessau
1737	12/23	Colonel Eugen, Prince of Anhalt-Dessau, later Major General
1744	3/10	Major General Christoph Ludwig von Stille
1753	4/13	Major General Georg Philipp Gottlob, Baron von Schönaich
1759	4/14	Major General Heinrich Rudolf Wilhelm von Vasold
1769	6/18	Colonel Just Rudolf von Seelhorst, later Major General
1779	1/15	Major General Theophil Ernst, Baron von Hoverbeck
1781	1/7	Colonel Hans Ludwig von Rohr, later Major General
1787	12/16	Colonel Carl August, reigning Duke of Saxe-Weimar, later Lieutenant General
1794	12/29	Major General Carl Wilhelm von Byern
1800	6/11	Colonel Christian Heinrich von Quitzow, later Major General

By the Elector's command of February 19, 1689 there arose from the four detached companies of Briquemault's mounted regiment (C.R. 5), with only 88 privates, a new mounted regiment under Major General Franz du Hamel, whose former regiment had been disbanded in 1679. It was quickly expanded to eight companies through recruitment in Westphalia so as to be ready for the campaign against France. After only two months, from April 16 to October 12 of that year, it took part in the tedious siege of Bonn, which was bombarded on and after July 24. In 1609 it joined the Prince of Waldeck's Corps to secure the area between the Rhine and Maas. In 1691 it had nine companies in the west again and endured the defeat of Leuze with the main army. At year's end it occupied the cities of Ghent, Ath and Namur under Heiden. It spent the winter at Cleves. In the same year it gave one company to the new Mounted Regiment 9, to be replaced quickly through recruitment in Halberstadt. In 1692 it was on occupation duty in Flanders, with no noteworthy events. In 1693 it was in the main army that was defeated at Neerwinden on July 29; it was hurt badly and lost its commander, Colonel Paul d'Ausson Villarnoul. In 1694 it was in Flanders, its strength unchanged, but marched to Magdeburg in the winter. The siege of Namur from July 17 to September 2, 1695 was the event of that year. The nineteen-year-old Prince Leopold of Anhalt-Dessau received his baptism of fire here: "Probably no man can imagine except he who from youth up has had so much desire to serve in his pilgrim heart", he wrote. In 1697 it was reduced to three companies for peacetime, but as early as 1699 it was built up to three squadrons again. At that time the regiment had five standards, about which its commander reported on March 5: "Standards are at hand that are made of the colors of cream and gold; on one side the Electoral eagle of Brandenburg and on the other side the hand of a clock which is held by an angel and an armored knight, with the motto: Ready every hour!—The drum banner is of the same colors and is marked with the Electoral eagle". In the spring of 1701 four companies joined Heiden's supply corps in the Wesel area. At the beginning of 1702 it sent twenty privates to Heiden's mounted regiment. Its chief went into the service of Venice after the King had refused to appoint him General of the Cavalry so as not to offend the Landgrave of Hesse-Homburg. So Colonel Charles, Count de l'Ostange took command and led two squadrons against Kaiserswerth, which was taken on June 15, a heap of rubble after an eight-week siege. Then it took Venloo on September 22, Roermond on October 7. In June of 1703 it was enlarged to six companies of 55 privates each. The year passed with the surrounding of Geldern under Lottum from April 21 to December 12. In 1704, its companies enlarged to 75 privates each, it went with Finckenstein to Rottweil to join the Anhalt-Dessau Corps on May 18. At Höchstädt on August 13 it captured two French flags but lost a standard when it broke through the enemy center in the second attack and overrode three infantry brigades. In mid-March of 1705 it marched to Lake Garda with the Dessauer, but saw no action in the unsuccessful action against the canals at Cassano. After it had lost almost all its horses to an epizootic, it turned back with the other two regiments and reached Halberstadt on foot in February of 1706. The cavalry regiments were replaced by infantry. As of April 12, 1709 it headed westward in the "New Corps" to take Tournay on July 29, its citadel on September 3, and to win at Malplaquet on September 3, capturing a pair of silver kettledrums. From May 2 to June 26, 1710 it besieged Douay, then Bethune and Aire, which were taken. In 1711 it took part in the capture of Bouchain on September 13, and in 1712 the siege of Landrecies, which began on July 16 and was broken off after two weeks. At the end of the war it came to Recklinghausen, Stift Essen, Werden and Bockum. The new King knew: "When one wants to direct something in this world, certainly the pen will not make it, if it cannot be asserted with a complete army". In the Pomeranian campaign in 1715 it helped to surround Wismar under von der Albe, which dragged on until April of 1716, causing great trouble for the troops. In 1718 it was strengthened to five squadrons.

Early in April of 1741 it joined the Observation Corps of Prince Leopold of Anhalt-Dessau in camp at Göttin on the Plane, south of Brandenburg/Havel; the corps broke up in October. From 1733 to 1806 its replacements came from the districts of Osterwiek, Halberstadt, Aschersleben and Oschersleben with their cities. Its garrisons were in Minden, Lübbecke, Rahden and Enger in 1714, in Mansfeld, Seehausen, Salze, Schönebeck and Walsleben since 1718, and from 1724 to 1806 in Aschersleben, Oschersleben and Kroppenstedt. Major General von Stille (1696-1752) was its chief for eight years, beginning in 1744; he was a highly educated, well-spoken officer, knowledgeable in military science, a teacher of Princes Heinrich and Ferdinand of Prussia, and outstanding cavalry leader at Hohenfriedeberg, and a welcome guest of the King. In the latter half of August, 1744 it marched to Bohemia in the First Corps and took part in the capture of Prague on September 16. In the privation-filled return to Silesia in terrible weather it lost so many horses that it had to be sent back to its garrisons in Magdeburg afterward. At the end of April 1745 it returned to Dessau's Corps at Megdeburg; it was ready to march as of July 27 and was twice strengthened late in August between Köthen and Dieskau. Ten days after the march into Saxony, Lehwaldt's Corps joined it on December 13. On the right wing at Kesselsdorf under Kyau, it led the successful enclosing attack from the east against the vigorously defended Kesselsdorf, opening the way for I.R. 30, so that the enemy position was rolled up. In the process it threw the Rutowsky Grenadiers back and took one of their flags, but lost three officers and 42 men.

In 1753 the King required that in attacks "the whole line must fall on the enemy's neck at once with all its power". In October of 1755 he had a new remounting of the cavalry carried out

6. Kürassier-Regiment
PRO GLORIA ET PATRIA
Uffz.
G. Dorn

outside the usual order, before arming began on June 19, 1756. On August 29 it joined the King's army to advance on Lobositz. There it rode with the left wing, which attacked without the King's orders, in the second attack south of the Lobosch. At Prague on May 6, 1757 it broke through the infantry at a gallop from the right flank in the last attack on Maleschitz, the squadrons one after another, against the enemy center and lost two officers and fifty men, in part from its own mistaken fire. At Kolin it formed the reserve behind the right wing under Bevern at Brzezan and held the retreat open at Planjan. After marching back with Bevern to the protection of Silesia via Moys and Barschdorf, it was west of Breslau on November22 facing enemy forces of three times its strength, whom the brave counterattacks under Kyau could not hold back. Taken to Parchwitz by Zieten to join the King on December 2, it attacked Nadasny's corps on the inner right wing at Leuthen and threw first its cavalry, then its grenadiers into flight from Gohlau to Lissa and on to the east bank of the Weistritz. The King expressed his "greatest satisfaction". In 1758 the cuirassier regiments were reduced by 120 horses. In the King's army it besieged Olmütz, returned to Silesia to serve under Margrave Carl as of August 10, and rejoined the King on September 11. At Hochkirch on October 14, in the saddle at the right time, it fought on the western edge of the village, cut I.R. 44 down completely, took 500 prisoners and captured a flag by charging forward repeatedly with entire squadrons in columns between the infantry. Then it helped cover the army's departure. In 1759 it was camped at Schmottseiffen with the King's army, leaving at the end of October to go to Saxony under Hülsen. Along with C.R. 7 and 9, it joined Finck's Corps on November 21 on the snow-covered high plains of Maxen, where they surrendered. In the mountain forest all attacks were unsuccessful. In spite of all its mobility, there was no escape. In Prince Heinrich's Corps in 1760 it joined C.R. 7 to form a complete regiment with five squadrons. The difficult reorganization took two years. At first it saw service only in the lesser actions in Pomerania, but in 1761 it was back in the King's army, fully formed along with its sister regiments, to advance into Poland, for example to Kosten, but then to go to camp at Bunzelwitz. In Silesia in 1762, strengthened to 1000 horses, it took part in the surrounding of Schweidnitz early in August, the encounter at Reichenbach on August 13, and the surrender of the fortress on October 10. After the War of the Bavarian Succession in 1778 and 1779, the campaign in The Netherlands in 1787 and that against the French Republic from 1792 to 1794, in which it saw action at Verdun, Valmy, Mainz, Saar and Rheinpfalz, it secured the demarcation line in Westphalia from 1796 to 1801. In 1806 three squadrons lost all but 95 of their men at Auerstedt. The regiment separated on the Uecker on October 28. One part surrendered in Anklam on November 1; what remained of the 4th and 5th Squadrons brought twelve officers and 250 men to East Prussia via Stettin and formed a squadron of the regiment that remained in existence.

6. Kürassier-Regiment

CUIRASSIER REGIMENT 7
Mounted Regiment

Commanders of the Regiment

1688	10/21	Colonel Friedrich Wilhelm, Baron von Wittenhorstzu Sonsfeld, later Lieutenant General
1711	6/19	Brigadier Georg Friedrich von der Albe, later Lieutenant General
1717	Jan.	Colonel Ludwig, Count von Wylich und Lottum, later Lieutenant General
1729	8/20	Colonel Carl Friedrich von Papstein, later Major General
1733	7/28	Colonel Friedrich Sigismund von Bredow, later General of the Cavalry
1755	7/3	Major General Georg Wilhelm von Driesen, later Lieutenant General
1758	11/22	Colonel Christian Sigismund von Horn, later Major General
1762	3/5	Colonel Leopold Sebastian von Manstein, later Major General
1777	8/13	Major General Gustav Ludwig von der Marwitz, later Lieutenant General and Commander of C.R. 2
1784	6/13	Colonel Friedrich Adolf von Kalckreuth, later Major General, Count since 1786
1788	7/11	Colonel Otto Friedrich von Ilow, later Major General
1792	11/12	Major General Hans Friedrich Heinrich von Borstell, later Lieutenant General
1804	12/3	Major General Heinrich August von Reitzenstein

As appointed on October 21, 1688, Colonel Friedrich Wilhelm, Baron von Wittenhorst zu Sonsfeld established a dragoon regiment, at first with one squadron, early in 1689; soon it was strengthened to a total of five with the addition of a company of Colonel von Perbandt. In the spring of 1689 it went to the lower Rhine with the army of Brandenburg to oppose France, and helped to enclose Bonn from April 16 to October 12, in order to protect the Rhineland and the army's connections. Meanwhile the mass of the regiment went to join Waldeck's army on the Mosel. It found winter quarters between the lower Rhine and Westphalia. In 1690 it went with Spaen's army to Brussels, joining Waldeck's beaten army at Fleurus on June 1 and remaining at its service for as long as needed. On March 4, 1691 it was strengthened by three ompanies, making a total of eight, but had to give up one for the new Brandt Dragoon Regiment (C.R. 11). Thus it had seven companies when, in Spanish service under Flemming, it protected the forts of Brabant, fought off French attacks at Liege from June 2 to 6, and moved into Luxembourg under esse-Cassel at the start of September in order to collect contributions. In 1692 it tasted defeat at Steenkerke on the Senne without being able to attack, and tried to relieve Charleroi. In 1693 three squadrons marched under Dewitz from Maastricht to Dren, then to Frankfurt and Heilbronn, and back to the lower Rhine in September. In 1694 seven companies advanced under Coehorn from Liege to Ghent and took part in the capture of Huy. As of July 2, 1695 it gathered at Visé under Heiden and took part in the siege of Namur until September 2. The Brandenburgers had "the greatest share in the conquest". The year of 1696 passed uneventfully. In 1697 it protected Cleves and Geldern from French raiders. In November it was reduced to three companies for peacetime and remained in the west, but on November 1, 1699 it was expanded to four companies again, having been transferred to Magdeburg at the end of 1698.

In April of 1701 its four companies belonged to Heiden's Corps and, under Prince Albrecht Friedrich, besieged Kaiserswerth from April 16 to June 15, took Venloo on September 22, Roermond on October 7, Liege on October 29, and Rheinberg on February 7, 1703. In June of that year it was enlarged to six companies. After helping to surround Geldern under Lottum from April 21 on, it went to Bonn, which fell on May 15, as did Huy on August 25. In 1704 it was sent to reinforce the Brandenburg Corps on the Danube, and on May 18 three squadrons joined Dessau's Corps at Rottweil. After the success of the French and Bavarian cavalry at Höchstädt, it led all the cavalry through the enemy center in the second attack of the decisive encounter,moving toward the Danube south of Blindheim. In the process it conquered a pair of silver kettle-drums. Prince Eugene praised the "unshaken steadfastness" of the Prussians who had helped to drive the French back across the Rhine. After the taking of Ulm on September 13 and Landau on November 23, it took up winter quarters in the Cham area. In mid-March of 1705 its eight companies with 75 privates each, 730 soldiers in all, marched under Stillen through Trient to Lake Garda. At Cassano on August 16 every attempt by the cavalry to cross the two Adda canals failed. At the end of October it had to start marching home on foot because of an epizootic among the horses; in February it reached the borders of Prussia at Halberstadt. Three infantry battalions replaced the absent cavalry in Italy. Soon after that it again joined Lottum's Corps with four squadrons and moved to the west, where 1707 passed uneventfully. In 1708 it was replaced in the corps for a time by the Bodyguard Dragoon Regiment. In 1709 it took part in the capture of Tournay on July 29 and the cavalry battle that completed the victory at Malplaquet on September 11. In 1710 its four squadrons were in Dessau's Corps at Douzy from May 4 to June 26, then at Bethune until August 29 and at Aire on the Lys until November 9, after an attack on the protective belt around Paris had failed at Arras. On June 19 Brigadier Georg Friedrich vonder Albe became chief of the regiment. In view of Prussia's defenselessness in the confusion of the Northern War, the Crown Prince wrote: "We have no regiments in the country, no powder, no money, and foreign troops whom we must treat like raw eggs. "Smoke-spewers" say they want to create land and people for the King with the pen. I say, with the sword, or he won't get anything!" In 1712 it served at Landrecies from July 16 on, until the siege was broken off after two weeks. When peace was made on April 11, 1713 it was in the state of Geldern with eight companies and 520 privates, and was soon strengthened by eighty to a total of 712 men.

Two squadrons left their camp at Stettin on May 1, 1715 and marched under Albe to Wismar, two others to Stralsund; while the latter city was being besieged as of October 20, one squadron took part in the Rügen landing from November 15 to 18. Then Stralsund fell on December 23. On June 11, 1717 Friedrich Wilhelm I made it and three dragoon regiments into "Regiments on Horseback" with four squadrons, increased to five on December 1, 1718. In 1719 it was garrisoned in cities, as was all the other cavalry in the land, after it had been in the Halberstadt area, and in Neidenburg and Lyck only in 1717 and 1718. Now it was garrisoned in Salzwedel, Tangermünde, Osterburg, Seehausen and varying smaller cities until 1806, with the exception of 1772 and 1773 when it was in Schwedt, Wriezen, Greifenhagen and Schönfliess. Replacements came from practically the same area from 1733 to 1806. In 1717 the King himself took a hand in the training of Albe's orphaned son. Its commander, Bredow, received the Order of the Black Eagle after Hohenfriedeberg. Its next commander, Driesen, once obligated to serve the King as a theology student, was an unusually talented, generally beloved cavalry leader who did a particularly fine job at Leuthen. In the spring of 1741 it went with the Third Corps to occupy Upper Silesia and northern Moravia. In the first half of May 1742 it went to the Chrudim camp in Bohemia. At Chotusitz on May 17 it was in the first line of the right wing,

7. Kürassier-Regiment
PRO GLORIA ET PATRIA
G. Dorn

storming to the south, throwing back the enemy cavalry, smashing the Croats and breaking into the second encounter behind the Austrian lines, whereby it attacked the infantry regiments of Palffy and de Vette and finally smashed a hussar regiment. The price was 132 dead, 168 wounded and 334 horses. The cavalry had more losses than the infantry that decided the battle. Two Pour-le-merite expressed the King's thanks. On July 25, 1744 the King commanded that the march onto the battlefield was to take place with closed squadrons in line, one after the other. The regiment belonged to the King's First Corps during the march into Bohemia that ended in the difficult return. After gathering at Wartha and Patschkau in April of 1745, it was on the inner left flank before Thomaswaldau at Hohenfriedeberg on June 4, making a unified attack under Nassau, first on the Saxon cavalry, then on the infantry regiment of Schönberg, later Count Brühl, which it rode down, losing five officers and 47 men. At the end of August it marched around Saxony with Gessler until October 6, when it reached the camp at Dieskau near Halle to join the Old Dessauer's Corps, which advanced to Kesselsdorf after one last enlargement on December 13. Here under Gessler in the center of the right flank, it pushed the enemy through the village and southern ravines and so far into the flank that its resistance was broken.

In 1756 the regiment reached Lobositz with the King's Army on October 1. In the second attack it swept the enemy cavalry from the field south of Lobositz, losing four officers, 33 men and 80 horses. At Prague in 1757 it was on the right wing in rough terrain and had only one last chance to save a victory after the breakthrough at Kej. While the battle raged at Kolin on June 18, it moved ahead to Brzezan to protect Manstein's flank and guard the rear. It suffered the third highest losses of the cavalry. On August 31 it marched westward to Rossbach, where it fought with distinction in two attacks of the second battle line. After a three-week march to Leuthen it rode on the left wing of the first line in Driesen's flank attack. In 1758 it was before Schweidnitz and Olmütz, then in the Saxon Corps. n June 24, 1759 it came to the Warthe under Hülsen and joined Dohna; on Jumy 23 it unselfishly attacked the Russian batteries on the Palzig Heights. In the second encounter at Kunersdorf, under Württemberg, it captured a battlement on the Spitz-Berg, losing thirteen officers and 136 men. In mid-September it went to Torgau with Finck. After the battles of Korbitz, Strehla and Pretzsch, the much-weakened regiment arrived at the support point of Maxen. In 1760 it joined C.R. 6 for a time to form a regiment in Prince Heinrich's Corps which, under Werner, carried out small actions in Pomerania. Reestablished at full strength in 1761, it joined the King's Army to carry out advances toward Posen. In 1762 it was increased to a thousand horses, took on protective duties at Burkersdorf and secured the siege of Schweidnitz until October 10.

In 1806 it lost two-thirds of its men at Auerstedt; Blücher gathered the survivors. The remainder capitulated at Magdeburg on November 11, while the depot troops passed through Danzig into East Prussia and later became C.R. 6.

7. Kürassier-Regiment

CUIRASSIER REGIMENT 8
Mounted Regiment

Commanders of the Regiment

1691	3/4	Colonel Christian Heinrich, Margrave of BayreuthKulmbach
1712		Colonel Albert Wolfgang, Margrave of BayreuthKulmbach
1716	June	Major General Stephan von Dewitz, later Lieutenant General
1723	4/15	Major General Friedrich von Egel, later Lieutenant General
1734	10/15	Colonel Friedrich Sigismund von Waldow
1742	5/19	Colonel Friedrich Wilhelm von Rochow, later Lieutenant General
1757	11/5	Lieutenant General Friedrich Wilhelm, Baron von Seydlitz, later General of the Cavalry
1774	6/8	Major General Maximilian Sigmund von Pannwitz, later Lieutenant General
1787	9/2	Major General Carl Friedrich Adam, Count von Schlitz, known as Görtz, later General of the Cavalry
1797	9/12	Colonel Ludwig Ferdinand Friedrich von Heising, later Lieutenant General

As ordered on March 4, 1691, a new regiment was set up for Margrave Christian Heinrich of Bayreuth-Kulmbach, uniting four companies of Lieutenant Colonel von Lehtmate, now Regimental Commander, the new company of the Bodyguard Regiment, and the new company of Captain Cormont of Bayreuth. The four original companies were made up half of companies recruited in 1689 to strengthened the Elector Prince Regiment (C.R. 2) and half of cavalrymen from Mecklenburg who had entered the service of Brandenburg. Including the staff and nineteen wagoneers, the regiment added up to 517 men and was intended for service in Hungary, to which it was already marching on May 4. On August 19 it fought in the bloody battle of Szlankamen, which wiped out the Turkish army; it lost 131 men. In October it stormed Grosswardein, then took up winter quarters in southern Slovakia, to march back to Brandenburg in May of 1692. Renewed in 1693, though only to 409 men—with 50 privates per company instead of 68—it went to Hungary, served before Semlin on August 9, and experienced the failed attempt to take Belgrade on September 7. It spent the winter in the Waag and March valleys. On September 19, 1694 it relieved the Imperial Army in its camp at Peterwardein and had to endure that army's frequent losses in 1695. As of May 18, 1696 it was in the field again with 401 men, taking part in the unsuccessful siege of Temesvar, but in 1697 it had an illustrious share in Prince Eugene's victory at Zenta on September 11, which laid the groundwork for the emergence of Austria-Hungary as a great power with the Peace of Carlowitz. In November of that year it was cut to three companies for peacetime, but enlarged to four companies again in November of 1698.

As of April 1701 it was part of Heiden's supply corps in the west, which assembled at Wesel. There it gave ten privates to fill Heiden's mounted regiment. In 1702 its two squadrons went to Kaiserswerth, which was surrendered in ruins on June 15; then Venloo fell on September 22, Roermond on October 7, Liege on October 29, and Rheinberg only on February 7, 1703. In June it was enlarged to three squadrons with companies of 55 privates each; on April 22 it marched under Lottum to Geldern, which was finally occupied on December 12 after being blockaded and bombarded. On May 18, 1704 it went to Rottweil to serve as a second reinforcement for Prince Leopold of Anhalt-Dessau's Corps on the Danube. Its companies were now filled with twenty privates added to each, and it marched with the Allied Army to Höchstädt. Here on August 13, after the enemy cavalry's dangerous advance, it rode against the center and right flank under Hesse-Kassel in that second attack of the Allied cavalry that broke through three brigades and threw the French back to the Danube at Blindheim. In the process it captured a standard. The taking of Ulm and Landau resulted. It took up winter quarters in the Upper Palatinate. In mid-June of 1705 it went with Arnim's Corps to join Marlborough on the Mosel. As of June 17 it left to join the Margrave of Baden in camp at Leuterburg, marched to Darmstadt and had to go on to the Brussels area and then to Lower Alsace, where it helped to take Hagenau. As of 1706 it had the job of securing the province, which was being misused by Sweden, Russia and Saxony as a thoroughfare area, and so was stationed in Prussia, the Neumark and Pomerania. In time of peace from 1713 on it was reduced to 390 privates, but soon later was enlarged to 480 privates, 36 junior officers and eighteen officers. In the 1715 Pomeranian campaign it was quartered in its canton as of April 1 and later secured the coast of Far Pomerania under Arnim, patrolling the coasts to watch for Swedish attempts to land, then went into winter quarters at the end of November.

On August 22, 1718 one squadron of Heiden's disbanded cavalry regiment, formerly Spaen's and founded in 1666, joined it. A new quartering plan moved it into cities as of 1719. For this the state, acording to the maintenance ordinance of March 1, 1721, had to pay "cavalry money" to the city treasury.

As of December 1, 1718 the four squadrons of 150 privates each were reorganized into five of 130 each. After they were enlarged to 132 men, the regiment had 31 officers, 60 junior officers, 685 privates and thirty supernumeraries, a total of 806. Its replacements came from the area around Allenburg until 1740, then from recruitment until 1747, and then until 1806 from the districts of Namslau, Strehlen and Nimptsch in Silesia, plus their cities, as well as Reichthal. Until 1717 it was stationed in Raglit and Tilsit, 1718 in the Neumark, until 1740 in Labiau, Wehlau, Allenburg and Goldap, in the Altmark in 1742, from 1743 in Ohlau plus Strehlen, Grottkau and Münsterberg/Silesia since 1746, since 1796 also in Löwen. Friedrich Wilhelm von Rochow (1689-1759), its chief for fifteen years, had served while a lieutenant colonel as Friedrich II's companion and mentor since his confirmation, and later proved himself very well as a cavalry leader at Chotusitz, Hohenfriedeberg, Soor and Kesselsdorf. Friedrich Wilhelm von Seydlitz (1721-1773)—Cornet in 1741, Captain in 1743, Major in 1745, Colonel in 1756 and General in 1757—was not only a capable leader but also very well educated, a teacher and chivalrous spirit, and interested in military science. He furthered the careers of such as Lossow and Groeling, and in 1763 in Ohlau, as "Inspector General", he took over the training and organization of the soon exemplary Silesian cavalry. In 1767 he became General of the Cavalry. For twenty years he was commander and chief. In 1741 the regiment was in the Old Dessauer's Observation Corps, camped at Göttin and Grüningen, from early April to the end of October. At the end of April 1742 it was ordered to go with the Dessauer's Corps to the Jgerndorf area, but its use in Moravia never took place. On May 9 it reached the Chrudim camp, from which the King advanced to Chotusitz on May 15. At the beginning of the battle it took part under Buddenbrock in a great attack of the right

8. Kürassier-Regiment

NON SOLI CEDIT
G. Dorn

wing in the first line, which came to a stop at the second line; it suffered many losses, including its commander, Colonel von Kortzfleisch. The regiment received two Pour-le-merite. In 1744 it went to Prague with the First Corps, the city surrendered on September 16. After the unfortunate return to Silesia in the fall it remained under Margrave Carl at Jägerndorf in May of 1745 and took part in the successful breakthrough at Neustadt on May 22, going on to camp at Frankenstein. In the first encounter at Hohenfriedeberg it took part in the attack of the left wing at Thomaswaldau, lost ninety men but received five Pour-le-merite. At Soor on September 30 it fought on the left wing in the middle of the first line, took 800 prisoners from I.R. 17 and 40, and captured ten flags, losing only 22 men. On November 23 it shattered the Saxons at Kath. Hennersdorf and captured three standards, three flags, four cannon and numerous prisoners. On December 13 it went with Lehwaldt's Corps to join the Dessauer at Meissen. At Kesselsdorf it was scarcely used in the middle of the left wing before the Zschoner Brook.

In 1756 the regiment rode in the second attack of the right wing at Lobositz, losing six officers, sixteen junior officers and 130 men. During the Battle of Prague it denied the enemy its retreat line to the Sazawa south of the city. At Kolin it took part in the exemplary breakthrough at Krzeczhorz under Seydlitz, with 806 men in the Krosigk Division. More than forty guns were taken. Seydlitz became a Major General and received the Pour-le-merite. On September 17 Seydlitz attacked the Imperial Army at Gotha and beat fivefold superior forces. On October 17 he relieved Berlin. At Rossbach Seydlitz, the youngest general, was given command of the cavalry; in the middle of the second battle line it rode in the attacks on the cavalry at Janus Hill and the rear of the enemy infantry at Tagewerben, capturing many cannon, five standards and two flags. The King promoted Seydlitz to Lieutenant General on the battlefield and gave him the regiment and the Order of the Black Eagle. On the right flank under Zieten at Leuthen it drove the enemy back to Lissa. In 1758, after the King's misfortunes at Olmtz, it marched to the Oder on August 11. Under Seydlitz at Zorndorf it smashed the victorious Russians in the Zabern Valley, destroying four battalions. It captured 37 guns, a standard and six flags, and lost 155 men. At Hochkirch it was on the left flank. On September 25, 1759 it distinguished itself at Hoyerswerda. In 1760 it was at Dresden, then at Liegnitz on August 15 it fought at the middle of the left wing and met the attack of three cavalry regiments, took five battalions of I.R. 1, 24 and 35 prisoner and captured eleven cannon and six flags while losing 111 men. At the head of Zieten's cavalry at Torgau it saved the victory in the evening. On February 15, 1761 it captured three Saxon battalions and three cannon at Langensalza and gathered contributions in Thuringia. Then it was camped at Bunzelwitz. In 1762 it took part in the battles of Burkersdorf-Leutmannsdorf and Reichenbach. During the war the regiment had not lost a standard. It consisted of 823 Prussians, 43 Saxons and 96 'foreigners'.

When Seydlitz died in 1773, all cavalry officers were ordered to wear mourning for two weeks. In 1806 it capitulated at Pasewalk on October 29, the depot troops at Schweidnitz on January 8; the remaining men escaped to East Prussia and later joined Cuirassier Regiment 1 in Breslau.

8. Kürassier-Regiment

CUIRASSIER REGIMENT 9
Mounted Regiment

Commanders of the Regiment

1691	3/4	Colonel Hans Ehrentreich von Schöning
1703		Colonel Philipp Ludwig, Baron von Canstein
1706	April	Colonel Hans Heinrich von Katte, later Field Marshal, Count since 1740
1741	6/5	Colonel Hermann, Count von Wartensleben, formerly Chief of C.R. 11
1741	11/28	Colonel Johann Adolf von Möllendorff, later of D.R. 10
1743	11/14	Major General Bernhard Heinrich von Bornstedt, later Lieutenant General
1751	9/5	Major General Johann Carl Friedrich, Prince of Schönaich-Carolath, later Lieutenant General
1758	1/5	Major General Jakob Friedrich von Bredow
1769	6/12	Colonel Friedrich Wilhelm von Podewils, later Major General
1784	9/19	Colonel Christian Friedrich von Braunschweig, later Major General
1787	10/14	Major General Johann Wilhelm von Manstein, later Lieutenant General
1797	1/23	Colonel Friedrich Jakob von Holtzendorff, later Lieutenant General

On March 4, 1691 Colonel Ehrentreich von Schöning received the authority to establish a cavalry regiment of six companies with a total of 517 men. Already-existing companies were brought together for the purpose, including Schöning's old company in the Elector Prince Regiment 2, one company from the Lüttwitz Regiment (founded 1666, made a part of C.R. 2 as von Thümen's Regiment), the new Ninth Company of C.R. 1 from Anhalt, and the new companies of Captains von Brandt and von Borck of C.R. 3. Just two months later the march to Hungary began, and on August 19 it met the Turks in the bloody battle of Szlankamen, along with C.R. 8. Margrave Ludwig of Baden wrote to the Emperor on January 17, 1692: "—according to the agreeing statements of all officers, with the help of the Electorate of Brandenburg Your Imperial Majesty's entire army would have been smashed to pieces". In October Grosswardein was taken, in May of 1692 the march home began. In mid-May of 1693 its six companies of fifty privates, 409 men in all, marched back to Hungary, joining the army on August 9 and taking part in the unsuccessful storming of Belgrade before going into winter quarters in southern Slovakia. In 1679 it only relieved the encircled camp of Peterwardein. After a string of Imperial losses at Lippa, Titel and Lugos in 1695, it was transferred back to Brandenburg in 1696 after the death of King Jan Sobieski, as the situation in Northern Europe was getting problematic. Thus it was spared Zenta. At first reduced to half its numbers for peacetime in 1697, it was strengthened to four companies again in November of 1698 because of threatening developments. The alliance with the Emperor, renewed on November 16, 1700, assured the recognition of kingly dignity "as soon as the Elector deigns to have himself proclaimed King of his Duchy of Prussia and crowned".

As of January 19, 1702 it went into the service of England and Holland to protect their sea power, under the command of Prince Albert Friedrich, the King's brother, with six companies and 437 men in all. Its four companies were strengthened with thirty men each from C.R. 2 and the Barfus Regiment, twenty from C.R. 1 and ten from C.R. 8. They joined the regiment on April 25 and 26 at Kaiserswerth, which had been surrounded since April 16 and surrendered on June 15. From Kaiserswerth it went across the Rhine from Dsseldorf in Heiden's Corps to besiege Venloo, which capitulated early, on September 22, by a lucky chance. On October 7 it took part in the capture of Roermond. Then it went via Cologne to join Marlborough in Brabant, besieged Liege and was able to take it on October 29. The conquest of Rheinberg under Prince Albert Friedrich took until February 7, 1703. Assisting the naval forces gave more and more reason for bitter complaints. Soon the regiment was ordered to take part in the siege of Bonn on the west side, under Marlborough. The city was captured on May 15. Then it fought at Maastricht. Soon Huy was enclosed, to fall on August 25. Otherwise the campaign in Brabant passed without great events. As of 1704 the regiment was under the Hereditary Prince of Hesse-Kassel, but was also put under foreign commanders at various times and hardly ever used as a unit, until in 1706 the regiment was finally put in foreign service in the west, along with the other Prussian troops, which were united under Lottum. It had done its duty under difficult conditions. On May 23, 1706 it took part in Marlborough's victory at Ramillies, north of Namur, where the French-Bavarian army was destroyed, with ten thousand prisoners, 54 guns and more than sixty flags taken. Then Louvain fell on May 25, Brussels on May 27, Ghent on June 1 and Ostend on July 6. As of July 11 it was sent to join the Lottum Corps, at its camp in Helchin on the Schelde, which besieged Menin from July 23 to August 23 in the presence of Crown Prince Friedrich Wilhelm. After taking Ath it spent the winter in Brussels. The year of 1707 brought few events other than marches between Soignies and Helchin. On July 6, 1708 Prince Eugene arrived, to lead it into battle at Oudenarde on July 11, where its two squadrons joined the attack on the right wing of the second line. Then the Lys crossings and Lille were taken, the Schelde line penetrated on November 27 and Ghent occupied. In 1709 it took part in the capture of Tournay on July 29 and rode in Natzmer's cavalry attack in the victory at Malplaquet on September 11. After helping to capture Mons on October 20, it went to Maastricht for the winter. In 1710 it took part in the capture of Douay on June 26, Bethune on August 29 and Aire on November 9, then spent another winter in Maastricht. In 1711 it was involved only in the surrender of Bouchain and spent the winter in Douay. After the taking of Le Quesnoy on July 4, 1712 brought the abandonment of the siege of Landrecies and the loss of almost all conquests. It spent the winter in Venloo.

In 1713 its six companies were built up into three squadrons of 75 privates each in place of 55. Summoned from East Prussia in 1715, it came out of its cantonal quarters to secure Far Pomerania against Swedish attempts to land, then went back to winter quarters at the end of November. On August 22, 1718 it was strengthened by the addition of a squadron of Heiden's Regiment, then divided into five squadrons on December 1. Until 1743 its replacements came from Angerburg, Barten, Lyck, Lötzen and Johannisburg, from Silesian infantry regiments until 1747, then from the districts of Oppeln and Falkenberg O/S with their cities and Proskau, Schurgast, Friedland and Krappitz. Its garrisons until 1740 were Angerburg, Barten, Lötzen, Lyck and Rastenburg in turn, in 1741 Lauenburg and Stolp, in 1742 Stendal, Gardelegen, Osterburg and Kalbe/Altmark, from 1743 on the canton, plus Löwen and Gross Strehlitz until 1755, since 1796 Neustadt/Oberschlesien. Its chief for 36 years, the subsequent Field Marshal von Katte, was the father of that unfortunate lieutenant who was executed in 1730 because of Friedrich II's well-known attempt to flee the country. From April to the end of October 1741 it was part of the Dessauer's Observation Corps, camped at Göttin and Grüningen, having been summoned from Far Pomerania. In 1742 it joined

9. Kürassier-Regiment

the Old Dessauer's Corps at Jägerndorf by the end of April, in order to march to northern Bohemia quickly because of the changed conditions. At Chotusitz on May 17, on the right wing in the first line, it took part in the Buddenbrock Division's victorious ride, which penetrated to the enemy's second battle line. In peacetime it was transferred to Silesia. In the King's First Corps in 1744, it experienced the withdrawal after the taking of Prague, caused by lack of supplies, bad weather and illness. Yet Winterfeldt wrote on April 25, 1745: "With our common man anything can be done that can be imagined, if only the officers want to give him help". At Hohenfriedberg on June 4 of that year it charged the enemy on the left wing at Thomaswaldau, led by Kyau, followed by Zieten and Nassau's Division. The King expressed his thanks with four Pour-le-merite. At Soor on September 30 it attacked south of Burkersdorf, shattered two enemy infantry regiments and captured 800 men and ten flags. On November 23 at Kath. Hennersdorf it attacked three Saxon cavalry regiments and one of infantry, taking 31 officers, including a general, and 1050 men prisoner.

In Schwerin's Corps in 1756 it led the feint from the County of Glatz to the Elbe. In 1757 it marched through Gitsch in-Jungbunzlauto Prague, where on May 6 it attacked under its chief, Lieutenant General Prince von Schönaich, on the left wing, moving from Unter Poczernitz to Sterbohol, breaking through to the second battle line, taking two standards and then having to retreat. It was spared Kolin. At the end of August it marched under Bevern to Silesia, where it took part in the Battle of Moys, the cannonade of Barschdorf and, on November 22, the fall of Breslau, which its attack under Kyau could not prevent. Under Driesen on the left wing at Leuthen, it overran the end of the Austrian right wing that was advancing toward Leuthen. In 1758 it besieged Schweidnitz with the King and moved to before Olmütz, whose siege was broken off early in August. It stayed in Silesia under Margrave Carl and reached the King at Grossenhain in September. At Hochkirch on October 14 under Zieten, it drove O'Donnell's cavalry back from the western edge of town and captured three standards. Afterward it marched via Görlitz to relieve Neisse. In 1759 it was in camp at Schmottseiffen with the King's Army, which Prince Heinrich took over. It marched with him to Sagan, then on to Bautzen, and late in September via Hoyerswerda to Strehla to join Finck's Corps, which was trapped and surrendered at Maxen. Its two newly organized squadrons were in Prince Heinrich's Corps at Landsberg, Glogau and on the Bartsch in 1760. At Primkenau it joined Goltz's Corps and was united for a time with three squadrons of Dragoon Regiment 8; then it returned to Silesia. In 1761, independent again, its three squadrons were at Strehlen, then in camp at Bunzelwitz and Pilzen near Schweidnitz, which fell on account of the retreat to Münsterberg. In 1762 it was filled to a thousand men again and protected the siege of Schweidnitz from August 8 to October 10, after the Battle of Burkersdorf.

After taking part in the War of the Bavarian Succession in 1778 and 1779, it saw action in the Polish campaign in southern Prussia and Poland. In 1806 it was disbanded with Hohenlohe's Corps in Pasewalk on October 29, the depot troop in the County of Glatz. The garrison troops joined the Silesian Cuirassier Regiment.

9. Kürassier-Regiment

CUIRASSIER REGIMENT 10
Gens d'armes

Commanders of the Regiment

1691	12/20	Colonel Dubislav Gneomar von Natzmer, later Field Marshal
1739	5/14	Colonel Wolf Adolf von Pannwitz, later Major General
1743	4/6	Colonel Georg Konrad, Baron von der Goltz, later Major General
1747	11/20	Major General Nikolaus Andreas von Katzler, former Chief of C.R. 3, later Lieutenant General
1761	4/9	Colonel Friedrich Albert von Schwerin, Count since 1762, later Major General
1768	6/24	Major General Hans Friedrich von Krusemarck, later Lieutenant General
1775	6/23	Major General Joachim Bernhard von Prittwitz und Gaffron, later General of the Cavalry
1793	11/5	Colonel Carl Friedrich von Elsner, later Lieutenant General

In November of 1687 two companies of "Grands Mosquetaires", each with 65 privates from noble Huguenot families and 22 servants, were organized in Prenzlau and Fürstenwalde as a "training school for young officers of the cavalry and dragoons", under Marshal Armand, Count Schonberg. In July and August of 1688 a third company of German nobles, under Lieutenant Colonel Dubislav Gneomar von Natzmer, was added. In 1689 they joined the campaign on the lower Rhine, under Schöning, and came into battle under their Commander, Christoph, Count zu Dohna, between Urdingen and Neuss on March 12, performing with notable bravery. Schöning reported: "I can give testimony of the Grand Mosquetaires that they showed their bravery splendidly and have earned Your Electoral Highness' most gracious esteem". Soon Kaiserswerth and Bonn were captured. They spent the winter in the Rhineland. In 1690, after crossing the Maas on July 22, they experienced the unfruitful campaign in Flanders. In 1691 they marched under Flemming to Namur, Charleroi and Huy. After the defeat of Leuze they joined in the advance on Luxembourg. By order of December 10-20 0f that year, the German Company of Colonel von Natzmer in Magdeburg and Halberstadt was changed into a squadron of two companies of "Gensdarmes", with all the privileges of the usual troops "of the Electoral House". Aside from the loss of Namur and the Battle of Steenkerke, in which the cavalry could not take part, 1692 consisted of marches and camps. In 1693 they left the County of Ravensburg under Dewitz and reached Maastricht, where they guarded the area between the Maas and Rhine under Flemming, then marched to join the Imperial Army in Heilbronn. In 1694 they were camped at Maaseyck under Dewitz, then in July they joined the main army and went to Louvain and across the Schelde to recapture Huy on September 27. They spent the winter in the Cologne area. In 1695 they set out from Visé to retake Namur from the southeast; it fell on September 2, and they spent the winter in the Imperial Abbey of Thorn at Roermond. Except for the march to Ath the year of 1696 remained uneventful. In May of 1697 they were north of Nivelles—the fighting had stopped—and went to Rousselaere, then to Ghent at the end of August, and came home in November. Reduced at first to one company of sixty privates, they were strengthened to two companies of forty privates each in November of 1698.

In July and August of 1700 they joined Brandt's Reconnaissance Corps at Lenzen on the Elbe. In 1702 they helped besiege Rheinberg under Prince Albert Friedrich, who led Heiden's Corps, until February 7, 1703, and secured East Prussia later that year. In 1704 the companies, now with 75 privates each, marched to the Danube in Dessau's Corps until August 3 and fought under Natzmer at Höchstädt on August 13, beating the French cavalry in the second attack and dispatching three infantry brigades. After that Ulm and Landau were taken. In mid-July of 1705 they met Marlborough on the Mosel, were at Lauterburg on July 25 and then went to Darmstadt, but had to return to Lower Alsace, where they took part in the capture of Hagenau on October 6. They secured East Prussia in 1706 and 1707. In June of 1708 they were in Lottum's Corps, which met Marlborough's army southwest of Brussels, where the Prussians were soon moved out of the second encounter to the head of the advance guard. At the victory of Oudenarde on July 11 General von Natzmer led the breakthrough between Müllem and Herlehem at the head of the Gensdarmes. When Ensign von Zieten sank from his horse, severely wounded, the standard was lost; Colonel von Canstein, the Commander, and eighty men fell. But they captured three flags and a standard from the Mosquetaires du Roi. Then they took part in the capture of Menin and Lille as well as Ghent. In 1709 they and the "New Corps" reinforced Lottum in the west and took Tournay. At Malplaquet on September 11 they fought in the center, under the eyes of the Crown Prince and Prince Leopold of Anhalt-Dessau, in the bloody, fluctuating cavalry battle, which cost the Prussians 143 lives. On October 20 Mons was taken. As of 1710 the Gensdarmes were recalled.

On July 1, 1713 the two companies of 80 privates were strengthened, becoming two squadrons of 150 privates each, and a third recruited in the Anhalt area for Prince Gustav of Anhalt, oldest son of Prince Leopold. A fourth came by command of December 30 of that year from the former Garde du Corps, which was reduced as of March 4 to one squadron of 150 men. From 1652 to 1692 they were called the "Satellite Guard" or "Mounted Bodyguard" They developed from the Bodyguard Company of the Burgsdorff Mounted Regiment, founded in 1633. Until 1740 the regiment was the only mounted guard unit. The King declared it the "first regiment of his house". The sergeants had the rank of ensigns in the army. In 1715 it was in camp at Stettin; one squadron landed on Rügen, and then it besieged Stralsund. On December 1, 1718 it was reorganized with five squadrons of 120 men each. Natzmer had developed the regiment for 48 years. He was a nobleman of high intellectual culture, organizing ability and strong leadership in dificult situations, though being a soldier was only a part of his life. He regarded himself as a "proper soldier" and a "true Christian" of earnest piety in the sense of A. H. Francke. He played an important part in the victories of Höchstädt, Oudenarde and Malplaquet, so that Friedrich Wilhelm I had good reason to name him Field Marshal. The regiment's replacements came from the districts of Halberstadt, Jerichow and the Priegnitz with Havelberg, Sandau, Hornburg and Stapelburg since 1733. The garrison was in Berlin, plus Bernau and Ketzin in1714, Rathenow and Wrietzen in 1716, with stables in Unter den Linden from 1779 to 1789, later near the State Library.

On December 16, 1740 one squadron, taken from the whole regiment and intended to protect the King, marched to Silesia on the left flank of the First Corps. The other four squadrons followed early in 1741 in the Third Corps. One squadron of 148

10. Kürassier-Regiment
NON SOLI CEDIT

men was set up in Berlin, then disbanded after the war. At Mollwitz the one squadron, under Schulenburg, was on the right flank under attack by threefold superior forces, and was driven back with the rest; it lost 36 men. In Schwerin's Corps the regiment went to northern Moravia on December 27 to besiege Olmütz. When it was attacked at night at Senitz, south of Wsetin, in March of 1742, half of it fought between the burning houses on foot until the others could mount and drive the enemy back with many losses. Early in May it marched to Bohemia under Derschau but only arrived after Chotusitz. In the First Corps in 1744 it experienced the surrender of Prague and the difficult autumn retreat. At Hohenfriedeberg in 1745 it rode down two Saxon battalions on the right flank, losing 42 men. At Soor it joined C.R. 1 on the right flank in von der Goltz's Brigade, driving fifty squadrons from the Graner Koppe and capturing 22 guns while losing only 68 men.

At Lobositz in 1756 it led Kyau's first attack on the right flank from the foot of the Homolka-Berg to open the battle, and captured two standards of the Cordova Cuirassier Regiment, losing 83 men including its commander. On October 16 it was on hand when the Saxons surrendered at Pirna. At Prague in 1757 under Keith, it enclosed the small side of the city during the battle, then led the retreat from Kolin to Leitmeritz to protect Saxony. As of August 31 it set out for Thuringia with the King. At Rossbach on November 5 it fought in the second battle line, first sending the caavalry flying and capturing seven standards, then slashing into the infantry on the other wing. Two Pour-le-merite expressed the King's thanks. A forced march lasting until December 2 brought it back to Parchwitz in Silesia to fight on the right wing at Leuthen under Lentulus, attacking Nadasdy's cavalry not far from Gohlau, overriding the Young-Modena Regiment and pursuing the enemy, capturing fifteen cannon, nine standards and five flags—and receiving ten Pour-le-merite.

In 1758 it experienced the siege of the Fortress of Schweidnitz and the fruitless enclosure of Olmütz, which the King lifted early in August. On August 11 it marched to the Oder and fought on the left wing at Zorndorf under Lentulus, bitterly attacking the advancing Russian wing across the Zaberngrund. It lost 98 soldiers but captured fourteen cannon and two flags, gaining four Pour-le-merite. Passing Küstrin and Dresden in mid-September, it arrived on October 14 to take part in the attack on Hochkirch, where its entire squadron advanced repeatedly between the infantry west of the village. "You couldn't get rid of them", an Austrian reported. After that it went with the King to relieve the Fortress of Neisse. In 1759 it gathered at Landeshut under the King and then went to camp at Schmottseiffen, then smashed the Vehla Corps at Hoyerswerda on September 25. In 1760 it took part in the luckless advance to Dresden, then fought at Liegnitz on Zieten's right wing which did not get into the battle, and on September 17 in the encounter at Hohgiersdorf and at Torgau, again in Zieten's Corps, whose attack from the south in the evening brought late relief. It spent 1761 in Silesia, camped at Bunzelwitz and Pilzen. Strengthened to 1000 horses in 1762, it was used at Adelsbach on July 6, then at Burkersdorf-Leutmannsdorf and on August 16 at Reichenbach, where it captured three standards and received a Pour-le-merite.

In the War of the Bavarian Succession it served at Jägerndorf in 1778. In 1794 it marched to Poland. In 1806 it fought at Auerstedt in the main army and went to Prenzlau in Hohenlohe's Corps. It was shattered at Wichmannsdorf on October 27; one part capitulated at Anklam on November 1, another escaped along with the depot to join the Stülpnagel Cuirassier Brigade in East Prussia.

10. Kürassier-Regiment

CUIRASSIER REGIMENT 11
Bodyguard Carabiniers

Commanders of the Regiment

1691	3/4	Colonel Paul von Brandt, later Major General
1692	5/10	Margrave Albert Friedrich of Brandenburg-Schwedt
1731	6/24	Major General Carl Ludwig, Count zu Waldburg, Hereditary Steward
1738	4/28	Colonel Hermann, Count von Wartensleben, later of C.R. 9
1741	6/5	Colonel Kaspar Ludwig von Bredow, later Lieutenant General
1751	11/14	Major General Peter Ernst de Pennavaire, later Lieutenant General, died 1/19/1759 of wounds at Breslau
1759	1/24	Major General Joachim Christian von Bandemer, died 1764
1768	9/26	Colonel Reinhold Friedrich, Baron von Hoverbeck, later Major General
1770	12/1	Colonel Reimar von Kleist, later Major General
1775	6/23	Colonel Philipp Christian von Bohlen, later of C.R. 1
1784	6/13	Colonel Heinrich Sebastian von Reppert, later Lieutenant General
1794	12/29	Major General August von Bismarck, later Lieutenant General, father of the Chancellor

After Friedrich Wilhelm came to the throne the companies were again expanded to 75 privates. In the Pomeranian campaign of 1715 it was camped at Stettin from May 1 on, then at Wollin in Arnim's Corps, which took the island of Usedom on July 31. Two squadrons landed on Rügen on November 15, after which it took part in the siege of Stralsund. On July 11, 1717 the King made it a "Regiment on Horseback", which was reformed in the Duchy of Magdeburg as of December 1, 1718 into five squadrons of 120 men each. On April 28, 1738, the day it came under the command of Colonel Hermann, Count von Wartensleben, the King raised it to a "Bodyguard Carabinier Regiment" without any particular privileges. The name was most probably taken from the disbanded mounted "Wartensleben Carabiniers", formerly the Barfus Horse, whose chief as of 1702 was Field Marshal Alexander Hermann, Baron von Wartensleben, General War Commissioner, since 1706 a Count of the Empire. As of 1733 its replacements came from parts of the districts of Jerichow, Zauche and Havelland with the cities of Neuhaldensleben, Wolmirstedt, Rathenow, Genthin, Jerichow and Burg. Garrisons were Landsberg/Warte, Fürstenfelde, Stargard and Werben near Colbatz in 1714, Rathenow, Burg, Wolmirstedt, Havelberg and Neuhaldensleben from 1716 to 1739, plus at first Möckern and Loburg, then Sandau and Genthin, as well as in peacetime and through 1806.

In 1740 it marched into Silesia with the First Corps under Duke Friedrich Wilhelm von Holstein-Beck; they had been mobilized as of November 25 and were to enclose Glogau. Glogau fell in a night attack on March 9, 1741, freeing the troops. On April 8 it crossed the Neisse with the King at Löwen-Michelau and went into battle at Mollwitz two days later. With Grenadier Battalion 5/21 in the center, it was on the right wing, where it took a direct hit from the Austrian attack and lost 135 men. Though it captured a standard, it was drawn into flight by its neighbors. The cavalry, as the King said, was "down so low that it believed I was delivering it to the slaughterhouse if I sent only a single detachment of it out". In months of time in camp it quietly practiced closeness and attack speed. The regiment's 300 men were involved in the siege of Olmütz by Schwerin on December 27, 1741. In February of 1742 it made an advance to Znaim with the King, and from May on it secured Upper Silesia. In 1744 it went to Bohemia with the King and experienced the failure of the campaign around Prague and the privations of the march back. At Hohenfriedeberg it fought on the right cavalry flank under Buddenbrock, came through the very fluctuating cavalry battle successfully, and then dispatched the Saxon Grenadier Corps, losing 57 men. Its Chief, Bredow, and two officers received the Pour-le-merite. At the end of August it passed through Sagan under Gessler and reached the Dessauer's camp at Dieskau, near Halle, on October 6. On December 15, after being reinforced, he ventured to do battle. On the outside of the right flank, it made a flank attack on Kesselsdorf and took the enemy's key point.

At Lobositz on October 1, 1756 it rode in the second attack of the right wing. The commander gained a Pour-le-merite. As of May 2, 1757 it enclosed the small side of Prague under Keith and cut off the enemy's retreat. In the Pennavaire Division at Kolin, it attacked the Starhemberg Division west of Brzistwi with 782 horses and was driven from the heights by overwhelming counterattacks, losing its Commander, Colonel von Schwerin. At the end of August it went to Silesia with Bevern, experiencing defeat at Breslau on November 22 that its vigorous attacks under the mortally wounded Lieutenant General von Pennavaire could not stave off. At Leuthen it led the flank attack of Driesen's left wing between Leuthen and Frobelwitz. In 1758 it took part in the siege of Olmütz. Attacked at Wischau, it defended itself bravely, and Colonel von Vasold received the Pour-le-merite. On August 11 the King marched to Frankfurt/Oder. At Zorndorf it was on the right flank under Schorlemmer with Seydlitz, beating back Demiku's major attack and helping the infantry move forward. Alle the staff officers were given the Pour-le-merite. At Hochkirch on October 14 it bitterly attacked O'Donnell's left wing and captured three standards. In 1759 it stayed in camp at Schmottseiffen until September 5, when it marched to Saxony under Prince Heinrich. Five companies under Colonel von Arnstedt were largely taken prisoner in an attack by General Luszinsky at Zeitz on February 17, 1760. Back in Silesia after the attack on Dresden had failed, it was on Zieten's inner right wing at Liegnitz but did not come into battle. On October 7 it set out via Berlin to Torgau, where it fought in Zieten's Corps in the evening and pulled out a victory with the last decisive counterattack at night. On February 15, 1761 it fought in Syburg's Corps under Colonel von Lölhöffel and captured two Saxon battalions and six grenadier companies at Langensalza. In the Saxon Corps, it then had to defend itself against greatly superior forces along the Mulde. In 1762, after the successful breakthrough across the Mulde on May 12, it again experienced nothing but small engagements until at Freiberg on October 29 the situation in Saxony was resolved. With a hundred men under the later Lieutenant General von Backhof in the Alt-Stutterheim Brigade, it took part in the attack on the Spittelwald. In 1763 it consisted of 769 Prussians, 27 Saxons and 75 'foreigners'.

On March 4, 1691 Colonel Paul von Brandt was given a dragoon regiment intended for Hungary, consisting of 409 men in five companies, and formed of one old company each of D.R. 1 of Ansbach, D.R. 2 of Sonsfeld, and the independent squadrons of Rauter and Perbandt, formerly of the Duke of Croy, and a new Perbandt company. It marched away on May 4 and took part in the advance on Belgrade as of August 3, which led to the bloody victory of Szlankamen on August 19. After the taking of Grosswardein and the march back in May of 1692, Margrave Albert Friedrich of Brandenburg-Schwedt was given the regiment, and the Rauter Company left. According to a renewed agreement of March 16, 1693 to provide a supply corps for

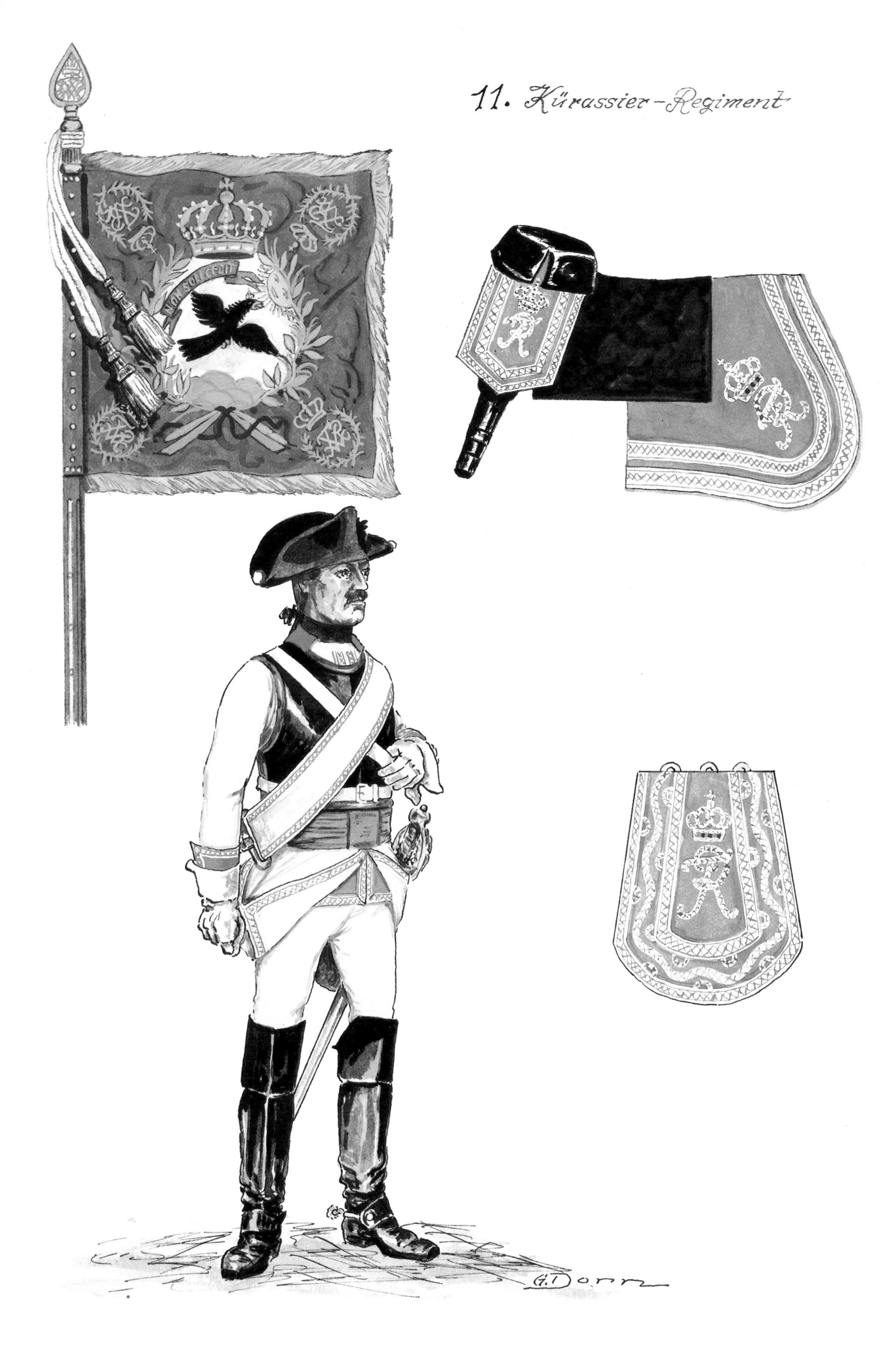
11. Kürassier-Regiment
NON SOLI CEDIT

Hungary, it again received a company of Ansbach's D.R. 1 and one of Rauter's Squadron that was replaced by a new formation, so that it could serve at Semlin on the Danube on August 9 with six companies of 50 privates each. After the unsuccessful storming of Belgrade it took up winter quarters in the Waag and March Valleys in October. In 1694 it relieved the Imperial army enclosed at Peterwardein. In 1695 it experienced the defeats of Lippa, Titel and Lugos without seeing action. It was called home to Brandenburg in 1696 because of reduction of the supply corps, not to mention annoyance, and to provide more security of its home area. Perhaps it was also to be spared, since its commander was the younger brother of the Elector Friedrich III, later the first King of Prussia.

When troop strengths were reduced in November of 1697 it was cut down to three companies like the other regiments, but was then given three companies of von Marwitz's Dragoon Regiment, formerly Derfflinger's, that had already seen action at Fehrbellin in 1675, in Pomerania in 1677 and East Prussia in 1679. . Thus it, like the Bodyguard Dragoon Regiment, had six companies in three squadrons and was transferred to East Prussia to secure the country in view of the coming Northern War (1700-1721). In 1703 its companies were strengthened to 60 privates each, in 1704 to 85, and two more companies joined it, so that it consisted of four squardons. When the war became more and more tempestuous and unpredictable in the neighboring countries, the troops in East Prussia were increased to seventeen battalions, nineteen squadrons. In February of 1705 it moved westward with a large portion of these troops to join the new corps under Arnim. In mid-June it made contact with Marlborough on the Mosel, but with an attack postponed to July 25, it was ordered to camp with the Imperial Army at Leuterburg. Called from there to The Netherlands, it received the command at Darmstadt to march back to Lower Alsace, where Hagenau was taken on October 6. In November it came to East Prussia again, under Arnim, for service there in 1706-1707. In the fall of 1707 the threat of war there diminished. Therefore, according to an agreement of March 31, 1708 it was transferred to Mecklenburg until the spring of 1709 to help Friedrich Wilhelm against his rebelling legislature and "bring the stubborn ones to reason". In 1710 it marched westward under Natzmer in the corps of Prince Leopold of Anhalt-Dessau to besiege Douay from May 4 to June 26, take Bethune on August 29 instead of Arras, and take Aire on November 9. It spent the winter in Limburg. In 1711 it fought between Douay and Arras from May on, broken only by the taking of Bouchain on September 13. In 1712 it took part in the capture of Le Quesnoy, while the siege of Landrecies had to be abandoned. From early November on it had winter quarters in Geldern. In January of 1713 it was in the Cleves area with 520 men, the companies reduced to 65 privates.

In 1806 it fought in the main army at Auerstedt, then in Hohenlohe's Corps, with which it surrendered at Pasewalk on October 29; some of its remaining men capitulated at Anklam on November 1, while one detachment and the depot reached East Prussia and later joined the Brandenburg Cuirassier Regiment.

11. Kürassier-Regiment

CUIRASSIER REGIMENT 12
Mounted Regiment

Commanders of the Regiment

1704	March	Senior Court Marshal August, Count Wittgenstein
1711	1/9	Major General Ludolf von Pannewitz, later Lieutenant General
1715	Nov.	Wilhelm Gustav, Hereditary Prince of Anhalt-Dessau, later Commander of C.R. 6
1715	11/30	Lieutenant General Benjamin Hieronimus Courold du Portail
1718	7/30	Colonel Georg Levin von Winterfeld, later Major General
1728	2/26	Colonel Arnold Christoph von Waldow, later Lieutenant General, died 4/3/1743 of wounds at Chotusitz
1743	4/7	Colonel Friedrich Wilhelm, Baron von Kyau, later Lieutenant General
1759	4/7	Major General Johann Heinrich Friedrich, Baron von Spaen
1763	2/9	Colonel Georg Ludwig von Dallwig, later General of the Cavalry
1796	10/2	Major General Jakob Friedrich von Berg
1798	3/1	Colonel Georg Christian von Werther, later Major General
1803	12/12	Major General Carl Wilhelm von Bünting

While all the dragoon regiments were being enlarged to eight companies in March of 1704, Senior Court Marshal August, Count Wittgenstein was given a new dragoon regiment at the same strength. Two companies of court dragoons, which he had taken over from his predecessor in the office, Count Lottum, formed the cadre. They had been assembled in August of 1689 by Senior Court Marshal Lieutenant Colonel Wilhelm Albrecht von Rauter for that purpose and strengthened in 1691, after other companies of the kind had existed in the past. They were joined by the two companies of Ordnance and Post Dragoons from East Prussia, which had been set up in 1674 for Königsberg-Warsaw communication, and had been commanded by Colonel Gotffried von Perbandt from February 15, 1684 to December of 1692. From them had come half of von Brandt's Dragoon Regiment, later C.R. 11, in 1691. One squadron hastened to the Maas and Lower Rhine, whither the regiment followed.

In February of 1705, its companies numbering 75 privates, it joined Arnim's Corps. In mid-June it reached Marlborough's army on the Mosel west of Trier. He was planning a great offensive toward France to avoid the ring of French fortresses, but gave up his intention because of their attacks in Belgium on June 17 and hurried there. The regiment, ordered to the Imperial army in Alsace, joined the Margrave of Baden in camp at Lauterburg, where the King's command to march to Belgium and join Marlborough arrived shortly afterward. Passing through Darmstadt, it got the counter-order to go back to the Imperial army, which it joined at Pfaffenhofen on September 15. Before it left for winter quarters in the Magdeburg area, it took part in the siege and capture of Hagenau from September 29 to October 6. Under Lottum in 1706 it went by way of Maastricht, Liege and Aalst to reach the Helchin camp on the Schelde in July, which Crown Prince Friedrich Wilhelm took an interest in visiting. One squadron took part in the siege of Menin from July 23 to August 23. After Ath had also been taken, it went into winter quarters between the Maas and Rhine. Unlike the successful year of 1706, 1707 went by almost without incident. the regiment advanced to Soignies, marched back to Tienen to protect Brabant, waited for results of Prince Eugene's break-through into Provence, then advanced to the Schelde again, this time to Helchin. The longest and hardest campaign developed under Prince Eugene in 1708. At Oudenarde on June 11 it marched up to the plateau of Heurne under Natzmer and then broke through on the right wing between Müllem and Herlehem. Then came the taking of Menin, the occupation of the Lys crossings and opening of the line from Comines to Ypres, the siege of Lille from August 22 to October 22, the breakthrough across the Schelde at Gavre and, at year's end, the recapturing of Ghent. In 1709 it was stationed in Prussia; in 1710 it joined the corps of Prince Leopold of Anhalt-Dessau that besieged Douay from May 4 to June 26, took Bethune, and conquered Aire from September 13 to November 9. It spent the winter in Limburg.

On January 9, 1711 Major General Ludolf von Pannewitz took over the regiment, after Count Wittgenstein had been deposed from his position in the "Three-Count Ministry" because of mismanagement. Since one company remained in Geldern and one in Wesel, it had three squadrons at Douay in early May and Bouchain in August and September. Comepete again in 1712, it took part in the campaign at Bouchain, Le Quesnoy and Landrecies, the siege of which had to be broken off without success; all the places that had been won were lost. On February 7, 1713, while in winter quarters, two companies were attacked in Odenkirchen near Rheydt by the French and 86 horses were lost. They came to East Prussia in July and gained sixty horses by absorbing Ciesielsky's Free Company—founded in April 1705 as postal dragoons—in August. Thus in July the regiment went to join du Trossel's Corps of the Imperial Army at Giessen and Philippsburg with three squadrons. In peacetime its eight companies had 75 privates each.

In 1715 it joined Arnim's Corps at Wollin for the Pomeranian campaign. To take Usedom, Arnim had the riders brought to a sandbank at the mouth of the Svine River at night, and the horses swam across. On the morning of July 31 he made a surprise landing of the regiment and 2000 infantrymen. Colonel von Preuss smashed the Swedish battalion and took 300 prisoners. Charles XII personally ordered the retreat; the infantry no longer attacked. The attack on Rügen could begin. "Because of proved exemplary bravery" the King promoted it to a "Regiment on Horseback" on August 6. On December 1, 1718 it was reformed with five squadrons. Until 1735 its replacements came from the Samland, until 1743 from Bartenstein and Heiligenbeil, until 1747 from recruiting in Silesia, then until 1806 from the district of Ratibor with Rybnik and Sohrau. Its garrisons were Rügenwalde, Lauenburg and Bütow as of 1714, Königsberg and Samland until 1740, Perleberg, Wittstock, Gransee and Kyritz as of 1742, Schweidnitz until 1746, Neustadt O/S, Oberglogau and Ziegenhals until 1752, then Ratibor and Leobschütz, plus alternating Katscher, Gleiwitz and Bauerwitz until 1806. Kyau (1708-1759), its chief for sixteen decisive years, possessed "uncommon knowledge of language and sciences through diligence and eagerness to learn". At the same time he was considered one of the most capable cavalry leaders at the start of the Seven Years' War.

From April to October 1741 it was in camp at Gttin and Grüningen with the Old Dessauer's Observation Corps, which was in the Jägerndorf area late in April 1742. After the withdrawal from Moravia, the regiment joined the King in camp at Chrudim. At Chotusitz it was on the left wing that stormed around the back of the enemy's second infantry encounter, where it disposed of two Hungarian regiments. In peacetime it stayed in Silesia. In 1744 it took part in the advance on Prague and the terrible march back, and at the end of May 1745 it was in camp at

12. Kürassier-Regiment
NON SOLI CEDIT
G. Donn

Reichenbach and Schweidnitz. At Hohenfriedeberg, after a long fight on the right wing before Pilgramshain, it swept the Saxon cavalry from the field. On September 30 it attacked south of the burning Burkersdorf in the direction of Soor. In Lehwaldt's Corps it reached the Dessauer's troops, who had been assembled at Halle since November 23, on December 13. At Kesselsdorf it was only prevented by the deep ravines of the Zschon Valley from fully smashing the Saxon army.

In Schwerin's Silesian Army in 1756, it took part in the feint from the County of Glatz, then marched through Jungbunzlau to Prague in 1757. Under Prince Schönaich, it attacked on the left wing over rough terrain south of Sterbohol on May 6; Zieten's flank attacks from the south decided the cavalry battle. At Kolin—with only 476 horses—it led the attack of Pennavaire's Division west of Brzistwi, moving uphill against Starhemberg's troops until it was thrown back to the Kaiserweg by the Saxons. In mid-August it went to Silesia under Bevern; under its Chief Kyau on November 22, its furious counterattacks could not hold its position on the Lohe. Three days after the King's return it was in the second line of the left wing at Leuthen, riding in Driesen's final attack, which he had waited for behind the Radaxdorf Heights. In 1758 it helped besiege Olmütz under the King from June 1 on after the taking of Schweidnitz. Under Zieten when Domstadtl was attacked on June 30, it was able to bring a large part of the ammunition and money train back to Troppau safely. After that it stayed in Silesia under Margrave Carl. With the King again north of Dresden as of September 11, it was on the left wing north of Rodewitz at Hochkirch and guarded the move back between Drehsa and Parschwitz. On November 7 it helped to relieve Neisse, and on November 20 it moved into Dresden. In 1759 it left the Saxon Corps under Hülsen to join Dohna's Corps on the Warthe. At Kay its attacks were defeated by the highland terrain and the enemy's strong counterattacks. It lost 260 men, including twelve officers, on the left flank at Kunersdorf, where it was hampered by the terrain and struck in the left flank by the enemy cavalry. It was at Görlitz from September on, then at Strehla. In 1760 it stayed with Prince Heinrich as he marched to the Warthe and northern Silesia. With the King as of August 29, it served at Wittenberg on October 23. At Torgau on November 3 it rode at the head of Holstein's riders and drove back the enemy cavalry, smashed the infantry regiments of Puebla and Wied, and struck four other regiments. Whole battalions were captured. It lost a standard but took two cannon and a flag. It was said: "The regiment had one of the leading shares in this victory. It was the first that attacked, and with so much good fortune that the first disorder was created". All staff officers received the Pour-le-merite plus 500 Taler apiece. It had lost more than half its men. In 1761 it was in camp at Bunzelwitz on September 9 to see the miracle of the Russian withdrawal. The King held off the opposition at Pilzen and Strehlen. Expanded to a thousand men in 1762, it secured the area along the Peile in the Battle of Leutmannsdorf on July 21, and guarded against attempts to relieve Schweidnitz at Reichenbach on August 13.

After the War of the Succession in 1778 and 1779 it distinguished itself in the Polish campaign of 1794, especially in the Battle of Boleslawice. In the main army in 1806, it capitulated at Pasewalk on October 29. The depot troops in Kosel and the men on garrison duty joined the Silesian Cuirassier Regiment.

12. Kürassier-Regiment

CUIRASSIER REGIMENT 13
Garde du Corps

Commanders of the Regiment

1740 11/16 Captain Otto Friedrich von Blumenthal, later Major
1744 5/10 Captain Georg Christoph Jaschinsky, later Colonel
1747 9/20 Captain Hans August von Blumenthal, later Colonel
1758 8/28 Lieutenant Colonel Wilhelm Diedrich von Wacknitz, later Colonel
1760 6/5 Major Carl Heinrich von Schaetzel, later Colonel
1773 12/23 Major Carl Friedrich, Baron von Mengden, later Major General
1785 9/25 Major Carl Wilhelm von Byern, later Major General
1794 12/30 Colonel Friedrich Wilhelm von Zollikofer
1798 10/2 Lieutenant Colonel Christian Friedrich von Rabenau, later Colonel
1801 5/1 Major Ernst August von Wintzingerode, later Lieutenant Colonel

The "Bodyguard Company on Horseback", formed in 1641, of the Jung-Burgsdorff Regiment of 1633, named "Bodyguards on Horseback" in 1652 and "Satellite Guard" in 1688, having two companies and 300 privates, were named "Garde du Corps" in 1692 and then had two squadrons; on June 1, 1713 they were merged with the Gens d'armes of C.R. 10. The old "Gardes du Corps" or "Old Satellites", only one company, were disbanded in 1708. One day after his father's funeral at Potsdam, King Friedrich II ordered the establishment of a "Garde du Corps" squadron, 166 men strong. Not only the officers, but the junior officers and privates as well were sought through transfers from all the cavalry regiments of the army. Formation took place in Charlottenburg during October and November; in November they were made ready to ride and put under the command of Staff Captain Otto Friedrich von Blumenthal. The officers' commissions were signed on November 16. Since the name meant nothing other than "Bodyguard", their chief was always His Majesty, the reigning King of Prussia, in whose name the commanders always directed them. Since 1440 the name had been used in France to refer to the king's escort troops who were on duty at court and in the field as far as the battlefield. In Prussia their uniforms were no more splendid than those of the army's other cuirassier regiments. But their standards were different, their cuirasses were silver and there were other minor differences. Their outer vests that they wore at court followed the styles of other courts in terms of color and the guard emblem, and thus were considerably different. From 1740 to 1806 they had no canton of their own, but received their annual replacements of three selected, experienced groups from the other regiments of the army, as was typical for elite units that were not mere court or parade troops but had the job of being true instructional and exemplary troops. Naturally their garrison was near the capital: in Charlottenburg as of 1740, Berlin and Charlottenburg from 1743 to 1752, in the barracks by the canal in Potsdam from 1753 to 1755; from 1764 to 1797 the First Squadron was in Potsdam, the Second in Berlin and the Third in Charlottenburg. From 1798 to 1806 the Guard, Second, Fifth and Ninth Companies were in Potsdam, the Fourth, Seventh and Eighth in Berlin, and the Third, Sixth and Tenth in Charlottenburg. The sergeants had the rank of First Lieutenant in the army.

In March of 1741 the squadron went into the field in Silesia with the C.R. 10 Gensdarmes and the rest of the Third Corps, at first more for securing than for fighting. In October of that year they, like the Guard Battalion, came home with the King and wintered in their garrison. On July 11, 1742 the First Silesian War was already over. In the latter half of August 1744 they marched through Saxony with the King's First Corps and reached Prague, which was taken on September 16 after a two-week siege. After the King had failed to achieve success in Bohemia, he tried at least to save his army in a winter withdrawal from that ravished land over the Silesian mountains. In December the squadron went back to Berlin with the Guards, and returned to Silesia in April of 1745. Now the King tried to decide the issue: "I have crossed the Rubicon and I want to maintain my power, or else everything shall fall". In the early morning of June 4 the Garde du Corps received their baptism of fire at Hohenfriedeberg, fighting on the right wing east of Pilgramshain. Along with C.R. 2 and 11 they set out across two deep ditches. Before the third a lieutenant and twenty dismounted riders set upa high, protective wooden fence eighty paces from the enemy. Then under Captain von Jaschinsky they put the Saxon cavalry to flight and smashed two battalions of the Grenadier Corps, losing six men and capturing seven standards and five flags. The Prussian cavalry, previously at a disadvantage, had proved to be "now equal to the opponent in mobility and daring". The King followed only as far as Königgrätz and waited for three months. On September 30 at Soor they fought on the right wing along with C.R. 2 and 12 and Dragoon Regiment 3, behind von der Goltz's Brigade, in the unexpected circling attack through a deep gully against the northeast slope of the Graner Koppe, and despite all the hindering terrain they drove two encounters and the reserves back to Altenbuch and Soor. In October they were already back in Berlin, which was threatened by the Saxons.

In 1756 they belonged to the King's First Corps, which marched into Saxony on August 29 and advanced into Bohemia in mid-September. At Lobositz on October 1 they rode on the right wing under Kyau in the first attack between the Homolka-Berg and Sullowitz—intended by the King as a show of force—and along with C.R. 2 and 10 they beat two enemy regiments and captured a standard of the Cordoba Cuirassier Regiment, losing 38 men. Lieutenant von Wacknitz took General Prince Lobkowitz prisoner. After the fall of Pirna on October 16, the King, who regarded the Saxons as "brethren in arms" and treated them accordingly, made two new squadrons out of the four in the Saxon Garde du Corps, founded before 1620, and promptly merged them with the Prussian squadron to form a regiment. But when it was said that not all of them "would fight against the enemy", Friedrich II had them dealt with on March 28, 1757, having each squadron keep nine junior officers, two trumpeters and forty men as reliable and exchange the rest for Prussians from the cuirassier and dragoon regiments.

At Prague on May 6, 1757 the First Squadron was in the right cavalry flank, which could scarcely take action in that terrain, the Second and Third Squadrons were west and south of the city, encircling it as far as the Sazawa. At Kolin only the First Battalion took part, staying with the reserves at the Kaiserweg. After the catastrophe the regiment returned to camp at Prague with the King and was used as guards. In mid-August it left Lusatia with the King and headed westward, leaving Dresden on August 31 and reaching Leipzig in the latter half of October. At Rossbach on November 5 it attacked on the right in the second line under Seydlitz, disaposed of the Austrian Brettbach and Trautmannsdorf Cuirassier Regiments and two French ones,

13. Kürassier-Regiment
„Garde du Corps"
PRO GLORIA ET PATRIA
Uffz.
G. Dorn

and captured four standards and two pairs of kettledrums. Pursuing the enemy to Eckartsburg, they took 800 prisoners, plus 32 officers, a general, three cannon and two howitzers. Back at Parchwitz in Silesia on December 2, they fought on the right wing in the Lentulus Brigade at Leuthen on December 5, first overcoming terrain problems and then breaking through both Austrian cavalry encounters at Gross-Gohlau, smashing the Jung-Modena Regiment and winning fifteen cannon, five standards and nine flags. Then Zieten's Hussars stormed forward from the third line. Nadasdy's Corps could no longer threaten the flank of the stormers at Leuthen, as they had been beaten off the field. In 1758 it marched with the King's Army to Schweidnitz, which fell on April 16, after two weeks, and to Olmütz, which was besieged as of June 1; there the enemy avoided any battle and yet forced the King to withdraw. On August 11 it left Landeshut with C.R. 8, 10 and 11 and went to the Oder to join Dohna's Corps. At Zorndorf it attacked on Seydlitz's left flank, under Lentulus in the Guard Brigade, which had been strengthened by Dragoon Regiment 4; after crossing the Zabern Valley they struck the forward-moving Russians and smashed their cavalry, then shattered five squares of infantry and captured fifteen guns and five flags. Captain von Wacknitz cried out: "A battle must not be considered lost until the Garde du Corps has attacked!" Then it led the counterattack of the right wing, still under Seydlitz, against the dangerous Russian cavalry attack under Demiku. After the battle Seydlitz reported to the King: "The Garde du Corps under Captain Wacknitz has done wonders!" The King promoted him to Lieutenant Colonel on the spot. The regiment had lost three officers and 43 men. On September 11 it was north of Dresden, and on October 7 it went to Bautzen, on the tenth to Hochkirch, where four days later, under Zieten, it furiously attacked the Austrians west of the village as they tried an encircling move, threw them back and thus protected the infantry's flank. In 1759 it was camped with the King at Schmottseiffen, where Prince Heinrich relieved him on Juny 29. After Kunersdorf it first guarded northern Silesia, then Lower Silesia. From July 10 to 19, 1760 it took part in the King's unsuccessful attempt to take Dresden and his famous seven-day quick march from the Elbe to Bunzlau. In the Battle of Liegnitz it was on the inner right wing, which did not get into battle. As of October 7 it marched with the King past Guben, reaching the Elbe at Wittenberg on October 23. At Torgau it rode at the end of Zieten's Corps, which waited at the Röhrgraben till darkness and then attacked the Süptitz Heights and inspired the last counterattack in the north. Here it lost 22 men. In 1761 it was present at Pilzen, Strehlen and in camp at Bunzelwitz, that "masterpiece of field fortification", where on September 9 it experienced the Russian withdrawal. In 1762 its strength remained unchanged although all cuirassier regiments were expanded. On July 22 it was used at the battle of Burkersdorf and Leutmannsdorf, and on August 16 at Reichenbach.

It took part in the War of the Bavarian Succession in 1778-1779 and the Polish campaign of 1794. Only on July 17, 1798 was it increased to five squadrons. In 1806 it fought in the main army at Auerstedt, went to East Prussia to join L'Estocq's Corps, and was taken over unchanged as the Garde du Corps Regiment until 1918.

13. Kürassier-Regiment „Garde du Corps"

DRAGOON REGIMENT 1

Commanders of the Regiment

1689	April	Margrave Georg Friedrich von Ansbach, died 3/29/1703 of wounds at Enshofen/Pfalz
1703-	1713	vacant
1714	2/27	Lieutenant General Andreas Reveillas du Veyne
1717	10/13	Colonel Georg Joachim von der Wense, later Major General
1725	8/13	Colonel Hans Friedrich von Platen, later Lieutenant General
1741	4/16	Colonel Carl Friedrich von Posadowsky, Count since 1743, later Lieutenant General
1747	4/12	Major General Berend Christian von Katte
1751	11/23	Major General Johann Ernst von Alemann
1755	5/30	Major General Carl Ludwig von Normann
1761	4/9	Colonel Johann Wenceslaus von Zastrow, later Major General
1774	6/26	Colonel Friedrich Albrecht Carl Hermann, Count von Wylich und Lottum, later General of the Cavalry
1794	12/29	Major General Ludwig, Prince of Prussia, died 1796
1797	2/9	Lieutenant General Ludwig, Duke of Pfalz-Zweibrücken, 1799 Elector of the Bavarian Palatinate, 1806 King of Bavaria

The oldest Prussian dragoon regiment became the source of a quarter of all dragoons. Three companies of the Imperial contingent in the war against France in 1689, under Margrave Georg Friedrich von Ansbach, were taken into the service of Brandenburg and, along with a fourth recruited in Cologne by Captain le Jeune, made into a squadron for the Margrave. As of the beginning of July it was already called to take part in the siege of Bonn. Meanwhile, under Schöning, it helped to fight off a French attempt to relieve Bonn via the Eifel Mountains, and for a time it was under the Prince of Waldeck. After Bonn fell on October 12, it spent the winter in the Rhineland. In 1690, after the defeat of Fleurus, it was transferred to Brussels, and joined Waldeck at Wavre on August 2. But nothing decisive happened. At year's end it secured the area west of the Maas. In 1691 the number of its companies rose to six, thus attaining the strength of a regiment. In the west it was able to prevent the loss of Liege. It spent August before Huy, where nothing happened. Early in September two squadrons, under Hesse-Kassel, joined the move to Luexmbourg to collect contributions. It spent the winter between the Rhine and the Maas. In 1692 it was at Tienen in the relief army for Namur, but the attempt was given up because of conditions on the Mehaigne. After spending time at Fleurus, it was called back to the Rhineland on July 5, was at Wavre again on October 23, and went under Heiden to join the troops in Spanish service. In 1693 it reached the main army at Louvain. In the defeat of Neerwinden on July 29 it put up brave resistance on the Windenbach between Laer and the Kleine Geete, then moved off to Diest. In 1694 it received two companies of ordnance dragoons from the squadron of the late Colonel von Rauter, under Captain von Dobeneck, but they stayed in East Prussia. At the end of July it reinforced the siege of Liege. On July 26 it was at Ghent, on September 27 it was present at the taking of Huy. In September two companies under Seckendorff went to the Cleves area to protect it from raiders. It spent 1695 at the siege of Namur, in the south bend of the Maas from June 19 to September 2. In the winter it was between the Rhine and the Weser. Nothing important happened in 1696, when it was at Louvain, Gembloux, Ath and Grammont. In 1697 it secured Cleves and Geldern from French raiders, against which the government called on the peasants to protect themselves. In November it was reduced to three companies; in 1698 it passed through Pomerania to East Prussia, where it regained its fourth company.

The Margrave of Ansbach died on March 29, 1703. In June it was enlarged to six companies, each with sixty privates in three squadrons, in 1704 to eight companies of 85 privates each. In 1706 it went through Liege and Ath in Lottum's Corps, reaching Helchin/Schelde in July; then one squadron took part in the attack on Menin, which fell on August 23. After the taking of Ath it went into winter quarters between the Rhine and the Maas. Nothing decisive happened in 1707, which it spent on the march. When Prince Eugene's army arrived in 1708, Marlborough tried to change things at Oudenarde on July 11. In breaking through the French lines between Müllem and Herlehem under Natzmer, it threw back the French cavalry on the left wing and captured silver kettledrums of the "Maison du Roi". The taking of Menin, the Lys crossings, Lille on October 22 and Ghent after the breach of the Schelde line on November 27 followed. In 1709 it besieged Tournay until July 29, then went to Mons. In the fluctuating cavalry battle of Malplaquet on September 11 it fought for victory under Natzmer despite heavy losses. Then Mons fell on October 20, 1710. Under Prince Leopold of Anhalt-Dessau it took part in the sieges of Douay from May 4 to June 26 and Aire on November 9. It spent the winter in Limburg and in 1711, after the advance to Lens, Bethune and Arras and the taking of Bouchain on September 13, it was transferred to the Altmark and Magdeburg on September 13 to protect the homeland. In 1713 the regiment had eight companies of 75 privates, making 600 in all, plus 24 officers, 56 junior officers, eight medics, eight flagsmiths, sixteen drummers and the regimental staff, 726 men in all. In the Pomeranian campaign of 1715 one squadron fought under von der Albe at Wismar, which fell only in April of 1716, and two squadrons took part in the landing on Rügen east of Putbus on November 15. At the end of November it went into winter quarters.

In 1717-1718 only two of the six old dragoon regiments existed, one of them D.R. 1, now under von der Wense. After it was reformed to five squadrons of 130 privates each in 1718, and had been given a light squadron of 150 privates in addition as of September 1, 1722, it was enlarged in 1725 to ten squadrons of 110 privates, and to 120 each in 1726. After the death of Major General von der Wense it was divided in Far Pomerania on August 13, 1725; one half remained under Colonel Hans Friedrich von Platen, the other went to Dragoon Regiment 2 under Colonel von Sonsfeld. The light squadron was doubled to 130 privates each, and in 1734 the squadrons were increased to 132 privates and increased the number of light squadrons to five, so that the regiment now had five heavy and five light squadrons. Its replacements came from the district of Rummelsburg and parts of the districts of Neustettin and Bütow with the cities of Belgard, Greifenhagen, Körlin, Tempelburg and Schlawe. Its Chief, von Alemann (1684-1757) had risen from the ranks. The garrisons were in the Altmark in 1714, in Rummelsburg, Polzin, Regenwalde, Köslin, Naugard and Neustettin as of 1716, and from 1746 to 1806 in Schwedt, Wriezen and Greifenhagen.

On December 29, 1740 it Marched into Silesia to besiege Glogau. On March 9, 1741 Lieutenant Colonel von Bornstedt stormed the walls with two squadrons and took some of the occupying troops prisoner. He received the Pour-le-merite. On the left flank at Mollwitz on April 10 it drove off the enemy in the first battle line, losing 86 men, and joined the infantry in the famous turn to win the victory. On April 16 the light squadrons, doubled since 1740, were separated to become D.R. 9 under Platen. The regiment was turned over to Colonel von Posadowsky. On December 27 two squadrons under Schwerin

1. *Dragoner-Regiment*

helped to take Olmütz; in February of 1742 the others came to Moravia with the King. In March it made a raid under Prince Dietrich via Ungarisch Brod and Messeritz to "four miles short of Vienna", was at Olmütz until April 23, then in Upper Silesia. In 1744 it went to Prague with the King and survived the march back. In 1745 it fought on the right wing under its chief at Hohenfriedeberg in the second line of the dragoons, against whom the attack of the Saxon cavalry broke before withdrawing after having been pushed far back. It lost 83 men. It was at Hirschberg during the Battle of Soor.

From September 10 to October 16, 1756 it took part in the bloodless siege of Pirna. At Herwigsdorf the Bodyguard Squadron lost 51 men. At year's end every squadron was increased by fifteen men. On April 21, 1757 it took part in the breakthrough at Reichenberg along with D.R. 12, lost 35 dead and 114 wounded, but captured three standards. Its chief, General von Normann, cut down Field Marshal Porporati himself. At Prague it attacked south of Sterbohol on the left wing in the second line of the cavalry battle, of which Zieten turned the tide, losing 81 men. At Kolin its 637 men led the victorious ride, unter Krosigk, against the unprotected flank of the Wied Division, captured five flags and forty guns. Afterward it smashed the Saxon Carabienier Guards in the back and took a standard. Eleven Pour-le-merite, the Grenadier Ride and three promotions were its reward, including Major von Platen's promotion to Colonel. While marching to Silesia under Bevern in mid-August, it fought at Moys, and then at Breslau on November 22, where it took many prisoners. On the right flank in the second encounter at Leuthen it attacked three Saxon and two Austrian cavalry regiments, took eleven officers and 800 men prisoner, and captured three standards. Then it threw two cuirassier regiments back and captured two Bavarian battalions with four flags. In pursuit of the enemy it captured two cannon. In 1758 it served at Schweidnitz and Olmtz, then marched to Zorndorf on August 11 where, under Schorlemmer on the right wing in the first line, it threw itself against Demiku's heavy attack, drove back the Second Battalion of I.R. 40 and a battery, helped Hussar Regiment 5 and completely smashed a Russian grenadier battalion. At Hochkirch it attacked on the western edge of town, for which it received the King's "Praise of the most outstanding bravery". At Ebersbach on October 26 it took eight officers and 418 men of the Austrian grenadier and carabinier corps alone prisoner. In 1759 it moved via Wobersnow to Posen and the Warthe, was in camp at Schmottseiffen and fought at Hoyerswerda on September 25 and then at Pretzsch. In 1760 it experienced the misfortunes of Dresden, did not see action in the center at Liegnitz on August 15, and fought at Torgau in Zieten's Corps. In 1761 it was at Pilzen and Strehlen and in camp at Bunzelwitz. Enlarged to 1000 men in 1762, it fought at Adelsbach on July 6 and protected the siege of Schweidnitz. In 1763 it had 867 men, including 53 Saxons and 344 'foreigners'.

In Weimar's Corps, then Blücher's, in 1806, parts of it capitulated at Mölln on November 5, Hansfeld on November 6 and Krempelsdorf on November 7, the mass of it at Lüneburg on November 12; the rest gathered in Danzig and East Prussia. In 1914 it was Dragoon Regiment 2.

1. Dragoner-Regiment

DRAGOON REGIMENT 2

Commanders of the Regiment

1725 8/13 Colonel Friedrich Otto, Baron von Sonsfeld, later Lieutenant General
1742 8/20 Colonel Louis, Prince of Württemberg
1749 5/14 Major General Reimar Julius von Schwerin, later Lieutenant General
1754 9/15 Major General Christian Friedrich von Blanckensee, died 1757 at Prague
1757 9/19 Colonel Anton von Krockow, later Lieutenant General
1778 11/8 Colonel Friedrich Wilhelm Carl, Prince of Württemberg, later Major General
1781 12/14 Colonel Johann Christoph von Mahlen, later Major General
1790 5/4 Colonel Carl Wilhelm Friedrich von Schmettau, later Lieutenant General
1797 1/23 Lieutenant General Wolf Moritz von Prittwitz

This regiment not only developed out of Dragoon Regiment 1 but also had many parallel experiences after it became independent in 1725. It was formed in 1689 out of the Imperial contingent from Ansbach, strengthened by recruiting, to be a squadron for Margrave Georg Friedrich of Ansbach. It was used immediately at Bonn but attained regimental strength only in 1691. It took part in the French-Dutch War in the west, but actually only until 1696, including the bloody defeat at Neerwinden in 1693. In 1697 it secured Cleves and Geldern, in 1698 Pomerania and particularly East Prussia. It entered the War of the Spanish Succession only in 1706—five years after it began—in Flanders, where it besieged Menin the same year, took part in the Battle of Oudenarde in the victorious year of 1708, and helped win the Battle of Malplaquet in 1709. Then came the fighting for the many well-built fortresses. In the autumn of 1711 it returned, though the war continued until 1714. In the Pomeranian campaign parts of it fought at Wismar, on Rügen and at Stralsund. In 1717-1718 it belonged to von der Wense's Dragoon Regiment (D.R. 1), which was enlarged from four squadrons with eight companies of 75 privates each to five squadrons with ten companies of 65 privates. In 1725 the division into companies was abolished and it was strengthened to ten squadrons of 110 privates each, increased to 120 a year later, and on August 13, 1725 it was divided. The new dragoon regiment was taken over by Colonel Friedrich Otto, Baron von Sonsfeld, who had been in the regiment since 1690. In 1734-1735 it took part in the Imperial Army's unsuccessful campaign on the middle and upper Rhine, in which Crown Prince Friedrich II drew some important conclusions about the condition of the Austrian troops, which surely had many consequences. King Friedrich Wilhelm I wanted the regiment to have ten squadrons and had it divided into ten companies as of August 19, 1739 and supplied with sufficient officers, so that with its staff and thirty supernumeraries it numbered 824 in all. This reformation was not completed during his reign. Its replacements came from the districts of Polzin, Belgard and parts of Greifenberg from 1733 to 1737, from then to 1742 from the area around Dinslaken, Rees and Duisburg, from 1743 to 1747 from Silesian infantry regiments, and from 1747 to 1806 from the district of Sprottau with the cities of Sprottau and Primkenau and the district of Freystadt with Freystadt, Beuthen/Oder, Neustadtel and Neusalz. Its garrisons were in Far Pomerania until 1719, in Tilsit, Ragnit, Pillkallen and Stallupönen until 1724, then until 1736 in Treptow/Rega, Wollin, Massow, Naugard and at times Polzin and Greifenhagen in turn, until 1740 in Duisburg, Dinslaken and Rees. As of 1742 it was in Silesia until 1777, in Bunzlau, Lüben, Raudten, then Haynau instead of Bunzlau, Polkwitz, and also Beuthen/Oder as of 1796.

In 1741 it was camped in Göttin with the Observation Corps of Prince Leopold of Anhalt-Dessau, which was moved to Grningen on September 12 and disbanded at the end of October. In 1742, when the Saxon threat had ended, it was between Rathenow, Burg, Genthin, Sandau and Neuhaldensleben. When peace was made it was transferred to Silesia. In 1744 it marched deep into Bohemia with the King's Corps; Prague fell on September 16. After successful action by the enemy, it set out in October on the unfortunate return march over the mountains to Silesia, in which the army lost almost all its strength. In 1745 it was at Jägerndorf under Margrave Carl, who broke out of a threatened encircling at the end of May and moved westward with 300 magazine wagons. Then the rear-guard was attacked on May 22 at Bratsch, not far from Neustadt, the regiment smashed the Imperial regiments of O'Gilvy (No. 46) and Esterhazy (No. 37) of over 1000 men and captured three flags, so that the dragoons' white coats looked "as if dyed with blood", as was reported. One officer and 62 men were dead, five officers and 88 men wounded. The column reached the Frankenstein camp at the right time. Its commander, Major General von Schwerin, and six officers, including Lieutenant Colonel von Alemann and Staff Captain Johann Dietrich von Manstein, received the Pour-le-merite. That was the way the King wanted his cavalry to be! At Hohenfriedeberg on June 4 it fought on the right wing next to Dragoon Regiment 1 before Pilgramshain, where the cuirassier encounter was first breached and the dragoons took the blow until the enemy cavalry withdrew and the grenadier corps was attacked, with the loss of one officer and sixteen men. At the end of June it went back to Upper Silesia with Nassau and proved itself at Neustadt on July 31, in the siege of Cosel, at Jägerndorf, Hultschin and Oderburg. When Prince Louis of Württemberg, its chief, entered French service in 1749, its commander moved up to the chief's position.

In 1756 it stayed with the Second Corps in Silesia under Schwerin and served only at Nachod from mid-September to the end of October. In 1757 it went over the Silesian mountains and through Gitschin and Jungbunzlau to Prague, where it reached the battlefield from the north on the morning of May 6. Here it entered the battle—as did Dragoon Regiment 1—on the left wing south of Sterbohol when the cuirassiers had to drop back. The turbulent great cavalry battle moved back and forth across the field many times, finally past Sterbohol on the south and eastward. Finally Zieten attacked the enemy in the flank and back and caused it to flee. Its chief, Major General von Blanckensee, was fatally wounded, and it lost 62 soldiers. It took many prisoners during the pursuit. On June 12 it fought under Bevern at Böhmisch-Brod. At Kolin a week later it attacked east of Krzeczhorz in Krosigk's cavalry reserve behind Dragoon Regiment 1 and broke through the enemy battle lines on the heights, almost the beginning of a victory. After the King had evacuated Bohemia as of July 21, it was at Görlitz on August 31, in order to march with Bevern via Tillendorf and Haynau to Breslau and take a position along the Lohe. On November 22 it fought on the left wing at Gabitz, south of the city. Zieten led Bevern's Corps to the King at Parchwitz on December 2, where the troops camped after a forced march from Leipzig. At

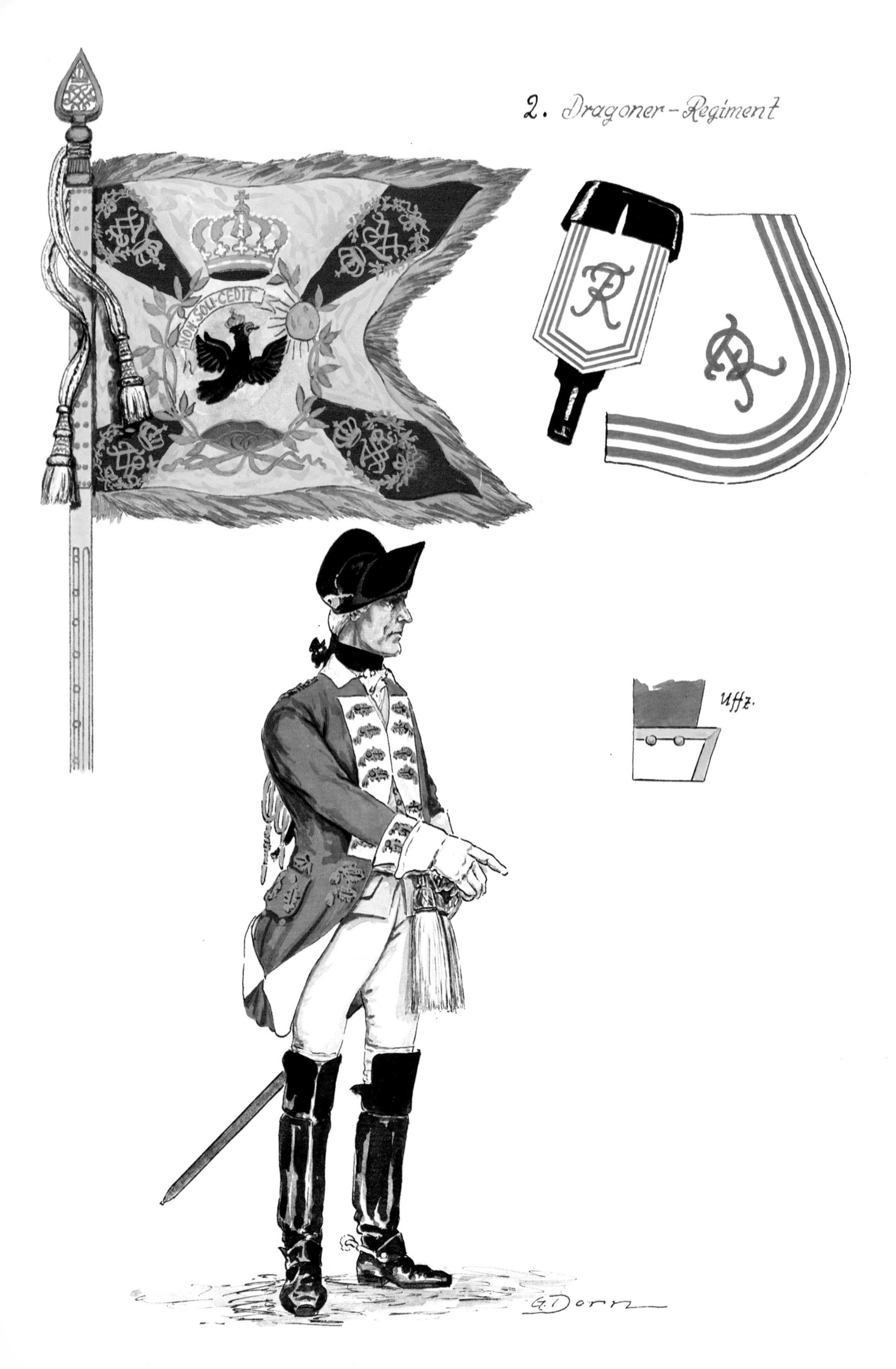
2. Dragoner-Regiment
NON SOLI CEDIT
Uffz.
G. Dorn

Leuthen it fought on the right wing under Zieten in the rugged terrain between Schriegwitz and the Weistritz, attacked Nadasdy's Corps and, again with Dragoon Regiment 1, drove it to flight after a fluctuating battle. When it then turned to take on two grenadier battalions, it was struck in the rear by enemy cavalry and lost 118 men. General von Krockow, its chief, was taken prisoner. With Zieten it pursued the enemy to Landeshut. In 1758 it marched with the King to Schweidnitz, which fell on April 16, and then to the siege of Olmütz until the end of June; after that was broken off it marched back to Königgrätz and advanced to Bischofswerda by the end of September before coming to Hochkirch; there it was on the left wing in the second battle line north of Rodewitz, fought off the attack and absorbed the retreat, losing two officers and 21 men. Then it followed the northern edge of the Sudeten Mountains to Neisse, relieving it on November 7. Ten days later it was at Glasen in Saxony, one day after that in Weissenburg, on November 20 at Dresden and on November 25 at Leibsch. In the King's Army again in 1759, it saw action successfully at Pfaffenholz near Reichhennersdorf on April 21, joined Prince Heinrich in July, reached Naumburg with the King on July 31, and at Markersdorf on August 2 under Kleist, attacked Hadik's baggage train south of Guben at night. The Mitzlaff Squadron captured a battalion of Blue-Würzburg with 23 officers and 800 men, plus two flags and three cannon. As of August 3 it was at Müllrose, not far from Frankfurt/Oder. At Kunersdorf it came onto the battlefield from the north, over the Trettin Heights, on the left flank of Finckh's Corps, to open the battle in Platen's Brigade. Despite the unfavorable terrain it moved forward spiritedly, making repeated vigorous attacks until it reached the Jewish Cemetery without any great success, losing almost two-thirds of its complement: eleven officers, forty non-commissioned officers, 484 men, 510 horses, but not a standard! Among the dead were its commander, Colonel Johann Dietrich von Manstein. This bloodletting was not without its consequences. Under Finck it successfully fought off superior enemy forces at Korbitz, near Meissen, on September 21. In 1760 it set up a light squadron with Polish remounts and took part in the unsuccessful attempt to take Dresden in July. At Liegnitz, alerted on the left wing in the morning twilight, it fell upon the enemy's flank and rear to gain time, beat every battle line and took 1000 prisoners and twelve cannon, three flags and two standards, losing 77 men and one standard. It received a gracious gift of 1000 Taler from the King. After the action at Hohgiersdorf on September 17, its 200 men lost 20 dead, 38 wounded and 86 prisoners when the Löwenstein Dragoons attacked it at Lindewiese. At Torgau—along with Dragoon Regiment 1—it fought in Zieten's Corps and helped to win victory late in the day. In 1761 it belonged to the Saxon Corps, which fought off the superior enemy forces on the Mulde in quiet but active defense and efficiency. In 1762 it was strengthened to 1000 horses and took part in the breakthrough across the Mulde on May 12, the success of which could not be maintained. After being hit hard at Brand on October 15, it advanced from St. Michael across the Spittelwald to Freiberg on October 11, along with D.R. 11 under Major General von Manstein in the Jung-Stutterheim Brigade.

In Hohenlohe's Corps in 1806, it capitulated with nineteen officers and 258 men at Prenzlau on October 28. The remaining men escaped to East Prussia, and the depot troops joined newly-formed units in Silesia.

2. Dragoner-Regiment

DRAGOON REGIMENT 3

Commanders of the Regiment

1704	12/10	Major General Friedrich, Baron von Derfflinger, later Lieutenant General
1724	2/2	Colonel Adolf Friedrich von der Schulenburg, Count since 1728, later Lieutenant General
1741	4/23	Colonel Friedrich Rudolf, Count von Rothenburg, later Lieutenant General
1752	2/11	Major General Georg Philipp Gottlob, Baron von Schönaich
1753	4/13	Major General Friedrich Ludwig, Count zu Waldburg, Hereditary Steward
1757	3/19	Major General Peter (von) Meinicke
1761	4/9	Major General Kurt Friedrich von Flanss
1763	4/8	Colonel Achaz Heinrich von Alvensleben
1777	8/13	Major General Otto Balthasar von Thun, later Lieutenant General
1788	6/4	Major General Georg Ludwig von Gilsa
1792	6/4	Major General Wolf Moritz von Prittwitz, later Lieutenant General
1797	1/23	Major General Hans Carl Ludolf von Strantz
1800	10/2	Colonel Friedrich Wilhelm von Irwing, later Major General

The Derfflinger Cavalry Regiments had lost their independence in 1697; that "of horse", founded in 1666, became Heyne's Regiment in C.R. 5, while the Dragoon Regiment, founded in 1655, became Marwitz's Regiment and the Dragoon Regiment of Margrave Albrecht in C.R. 11. The son of the Field Marshal, Major General Friedrich, Baron von Derfflinger, received orders on December 30, 1704 and January 3, 1705 to form a new dragoon regiment of eight companies in four squadrons. The five existing dragoon regiments provided ten men per company for it, 400 in all, and 75 men per company were recruited. From 1705 to 1708 it was stationed in East Prussia for protection during the Northern War, under Duke Friedrich Ludwig of Holstein-Beck. As of April 12, 1709 it went into the "New Corps" as reinforcement for Lottum in Flanders, where the War of the Spanish Succession was going on. The Allies attacked Tournay instead of lille, and the city fell on July 29, the citadel on September 3. On September 11 it fought at Malplaquet, the bloodiest battle of the war, in the alternating cavalry attacks when the French cavalry attack had come to a stop against Finckenstein's battlements. On October 20 Mons was taken. In 1710 it marched back to Prussia with the Gensdarmes. In 1711 it came back to see action there, having been exchanged with the Wartensleben Mounted Regiment. With Marlborough it went to Lens, Bethune and Lillers in mid-June. After the taking of Bouchain on September 13 the year was over. Although the English left in May of 1712, Le Quesnoy was taken on July 4 and Landrecies besieged as of June 16, to be broken off after reverses at Denain and Marchiennes. In the end the campaign's gains were lost. With its eight companies of 520 privates it spent the winter in Geldern and then in Ravensberg.

In 1713 the companies were strengthened to 75 privates again. On August 22, 1714 King Friedrich Wilhelm I honored the regiment by naming it "Mounted Grenadiers", "because it has shown so much distinction". This was unique in the Prussian army. Presumably it had done very well at Malplaquet, which is rather unlikely, by imitating either French or Austrian regiments; "Grenadier" served as an honor. Since then the regiment wore grenadier caps. On the hundredth birthday of Kaiser Wilhelm I on March 22, 1897, Kaiser Wilhelm II gave this name to Dragoon Regiment 3. Counting the staff, it had just 726 men. Ordered to camp at Schwedt on April 10, 1715, it was at Stettin on May 1. In the latter half of June one squadron went to the siege of Wismar under von der Albe. On June 28 the Peepe was crossed. Two squadrons took part in the landing on Rügen under Prince Leopold of Anhalt-Dessau on November 15. In the attack on the Swedes the regiment lost its commander, Colonel Otto, Count zu Waldburg; the Swedes surrendered on November 16. The regiment went into winter quarters at the end of November.

In 1718 it and Dragoon Regiment 1 were the only remaining regiments stationed in the Kurmark. General von Derfflinger (1663-1724) had studied in Frankfurt and Tübingen, taken extensive trips in Europe and first served in the Republic of Venice. He was succeeded in 1724 by Colonel Adolf Friedrich von der Schulenburg-Beetzendorff (1687-1741), a man very highly esteemed by the King, who had been in Prussian service since 1713 after studying for three years at Utrecht, and often sent on secret diplomatic missions to Hannover and London. He fell at Mollwitz, the first of the sixty Prussian generals to fall who remained in 1763. In 1724 the regiment was reformed with five squadrons of 130 privates each, then doubled in 1725 to ten squadrons, without division into companies, of 110 men each, increased to 120 in 1726. Its planned use against Hannover in 1729 did not take place. Plans to use it for the campaign on the upper Rhine in 1734 came to nothing because the supply corps was cut to ten thousand men. On October 1 of that year it was strengthened by ninety men. As of 1733 its replacements came from parts of the district of Landsberg/Warthe, Friedeberg and Arnswalde, later also from Meseritz with the cities of Dramburg, Reetz, Bärwalde, Lippehne and Friedeberg/Neumark. Its garrisons were Landsberg/Warthe, Wriezen, Briest and, as of 1720, Arnswalde, Guoritz and Woldenberg in rotation until 1725, then until 1739 Küstrin, Bärwalde, Dramburg and Friedeberg/Neumark, from 1746 to 1755 Neudamm, Küstrin and Bärwalde, then Friedeberg/Neumark, Driesen, Arnswalde and Berlinchen until 1806.

On December 16, 1740 it was on the right wing of the First Corps for the surprising march into Silesia. At Mollwitz on April 10 four of its squadrons were in each of the first two battle lines of the right flank, under its chief. When the battle began, it was struck by threefold superior forces, losing eleven officers and 255 men. When Count Schulenburg led the four squadrons of the second line against the enemy in vain, a standard was captured, but he was killed. The regiment was divided as early as April 21: Squadrons 1, 5, 6, 7 and 10 stayed in the regiment under Colonel the Count von Rothenburg, while 2, 3, 4, 8 and 9 formed the new Dragoon Regiment 4 under Colonel von Bissing. As ordered on September 4, the grenadier caps were replaced by hats. Without being removed, the name "Mounted Grenadiers" disappeared. The reason was the failure of the two escort squadrons of the regiment in the hussar attack at Baumgarten south of Frankenstein on February 27, from which the King barely escaped, but which had cost prisoners and a standard. In his instructions of March 3 and 23 the King had referred to that. In October it went with Hereditary Prince Leopold to the upper Elbe in Bohemia, to join the King's Corpsat Olmütz in February of 1742 for the advance to southern Bohemia, which was given up early in April. Under Buddenbrock at Chotusitz on May 17, on the right wing in the second battle line it beat first the cavalry and then the infantry to open the battle, until it had to give way

3. Dragoner-Regiment
NON·SOLI·CEDIT
G. Dorn

to Hungarian hussars. Its losses included six officers. In 1744 it marched in the King's First Corps to Prague, which was taken on September 16. Outmaneuvered in Bohemia, the army began to march home. The regiment, sent to Prague at the last minute, fought its way through via Leitmeritz to Friedland. It was sent to Küstrin for refreshment. On the left wing in the second line under Nassau at Hohenfriedeberg, it attacked successfully before Thomaswaldau, losing 36 men. At Soor it led the difficult cavalry attack, under Buddenbrock, on the northeast slope of the Graner Koppe, key to the enemy position; the whole wing was driven from the field, but the regiment lost only eighteen men and 59 horses.

In the King's Army on August 29, 1756 it crossed the Saxon border to fight at Lobositz north of the Homolka-Berg on October 1, riding in the unordered second attack that swept the enemy cavalry off the battlefield again. It lost fifty men. On October 16 the Saxons surrendered at Pirna. In 1757 it was on the right wing at Prague, where the terrain made fighting very difficult. At Kolin on June 18 it was among the reserves on the Kaiserweg. At the end of the battle it unselfishly attacked the pursuers of Tresckow's Division and Manstein's unit eight times to keep the retreat open. Then it moved westward with the King via Leitmeritz and Bautzen. On September 17 it, Dragoon Regiment 4 and Hussar Regiment 1 under Seydlitz attacked the Imperial army in Gotha without infantry and smashed it, though it outnumbered them fivefold. It had already fought at Rossberg with Dragoon Regiment 4 and the cuirassier regiments of Bretlach and Trautmannsdorf before the cuirassiers of the second battle line arrived. The second attack completed the victory. On the march back to the east it advanced from Chemnitz to Leitmeritz and destroyed the magazine. In 1758 it joined the Saxon Corps; on December 1 at Hornberg it caused the French to lose 400 men, and it had fought at Schladen, Hildesheim and Eldagsen by early March. In May it attacked in Franconia under Driesen; in August it defended at Pirna; at the end of the year it did the same at Landeshut. Under Belling at Sebastiansberg on April 15, 1759 it smashed three battalions and took 1300 prisoners, three flags and three cannon. On May 11 at Himmelkron, near Kulmbach, it captured the Kroneck Battalion, 21 officers, 522 dragoons from the Pfalz Regiment, three flags, two standards and three cannon while led by its chief, General Meinicke, and then advanced on Bamberg. Via Bautzen and the Schmottseiffen camp it moved to Kunersdorf, where it attacked in vain on the left wing in the first line at the Kuhgrund and Deep Path, losing eleven officers and 148 men. The enemy cavalry cut the infantry down. Afterward it went to Saxony. In 1760 it was in Prince Heinrich's Corps at Sagan, Glogau and Breslau, fighting against the Russians, then going to join the King's Army at Liegnitz. From Hohgiersdorf it moved westward to Wittenberg. Back in Silesia with Goltz's Corps, it was led by Zieten to the camp at Bunzelwitz. After the Russians withdrew, it joined Platen's Corps to attack the Gostyn camp on September 15. Then it went to Far Pomerania via Landsberg. At year's end it was in Saxony. In 1762 its 1000 men advanced on Teschen in Werner's Corps and were at Reichenbach on August 13. On August 16 it caused the Archduke Joseph Dragoon Regiment 200 losses and captured three standards.

In the main army in 1806, it capitulated at Ratekau on November 7, down to 160 men. In 1914 it was called Mounted Grenadier Regiment Baron von Derfflinger (Neumarkish) No. 3.

3. Dragoner-Regiment

DRAGOON REGIMENT 4

Commanders of the Regiment

1741	4/23	Colonel Wilhelm Ludwig von Bissing
1742	1/2	Colonel Friedrich Wilhelm, Baron von Kannenberg, discharged because of wounds
1742	8/18	Lieutenant General Carl Ludwig von Spiegel, died 10/19/1742
1743	1/3	Colonel Casimir Wedig von Bonin, later Lieutenant General
1752	9/21	Major General Henning Ernst von Oertzen, died at Lobositz
1756	10/4	Major General Carl Emil von Katte, died 11/16/1757
1757	10/24	Colonel Ernst Heinrich, Baron von Czettritz, later Lieutenant General
1772	9/22	Colonel Georg Ludolf von Wulffen, later Major General
1782	9/21	Major General Carl Ludwig von Knobelsdorff
1786	5/29	Colonel Carl Ludwig von Götzen, later Major General
1789	1/31	Colonel Georg Balthasar von Normann, later Major General
1792	11/12	Major General Friedrich Heinrich von Katte, later Lieutenant General

This regiment grew from the same roots as did Dragoon Regiment 3 and followed the same course until 1741. After the young Kingdom of Prussia had taken part in the War of the Spanish Succession for three years with a supply corps in Belgium, on the upper Rhine and in Italy, the regiment was organized on January 3, 1705 for the son of Field Marshal Georg von Derfflinger (1606-1695), creator of the army of Brandenburg, whose regiments had been disbanded two years after his death, and who played a somewhat unusual role. As an Austrian on Bohemian and Swedish field duty he rose quickly, then joined the service of Brandenburg in 1655 as an organizer and outstanding troop leader, and became a field marshal in 1670, a baron of the Empire in 1674; he was a towering personality. The regiment generally was given special consideration; it spent its first four years on securing duty in East Prussia, and was called back in 1710 and 1713. Its high point was its use in the west at the Battle of Malplaquet on September 11, 1709, while 1711 and 1712 brought marching maneuvers and a succession of sieges of small fortresses. As the supply corps of a coalition army, generally on subsidies or foreign pay, it was nevertheless not easy. On the other hand, the war supported the army and provided experience, even though the heart of its country was unprotected and became more and more of a thoroughfare for foreign powers. Appearance, order, reliability and preparedness for fighting were always stressed and praised in the Prussian regiments, unlike those of the allies. The strength of the companies varied between 65 and 75 privates, depending on the situation. The title of "Mounted Grenadiers" was extraordinary only for Prussia, but not inherently. It is still not clear why this regiment was chosen, rather than another capable one, as its achievements to that time were not exceptional. It may perhaps have been a gesture to General von Derfflinger by the new King Friedrich Wilhelm I, as a living link to his famous father. In 1714 the regiment had four squadrons, eight companies and a total of 24 officers, 56 non-commissioned officers, eight medics, eight flagsmiths, sixteen drummers and 600 privates, making 712 men, a total of 726 counting the regimental staff of fourteen.

It was transferred to the Lenzen camp in November of 1713, where a concentration of troops was meant to put pressure on Denmark for the benefit of the Duke of Holstein, but no action was to be taken. In June of 1714 Prussia allied itself with Russia, Hannover, Saxony and Denmark against Sweden, in hopes of finally acquiring Stettin and the land from the Oder islands to the Peene. On February 23, 1715 the Swedes opened the war, although they were inferior in numbers, even though they could depend on coastal fortifications, especially Stralsund. The taking of Usedom and the Peenemünde battlements by Arnim was an achievement; the landing on Rügen with a corps of 24 battalions and 35 squadrons from Prussia, Saxony and Denmark, departing from Greifswald, was a master stroke. Prince Leopold of Anhalt-Dessau said: "It will not be necessary to praise their bravery, because they are all honest, good people . . . A retreat will not have to be considered, and this must particularly be impressed on the rank and file". The transport fleet could leave only twelve days after boarding, and was at sea three days because of frequent opposing winds that November. The regiment took part successfully. As of 1718 it, like Dragoon Regiment 1, became the source of other regiments, since the dragoons had to be expanded again. Thus in 1725 it was enlarged to ten squadrons, even though the numbers of men were lower, but these were raised step by step. The time of the actual division remained open. Replacements came from the neighboring canton of Dragoon Regiment 3: the districts of Landsberg/Warthe, Friedeberg/Neumark and Dramburg with the cities of Landsberg, Arnswalde, Woldenberg and Schönfliess from 1741 to 1806. Its garrisons were Landsberg, Friedeberg and Arnswalde as of 1743, Woldenburg too as of 1753, just Landsberg and Woldenberg as of 1764 and Bärwalde as well since 1796.

Its participation in the march into Silesia on December 16, 1740 can be regarded asa sign of its worth. Its failure at Mollwitz indicated the quality of cavalry at that time, its misfortune at Baumgarten showed its sluggishness. The cavalry was not to let itself be thrown into disorder by attacks by swarms of hussars and was to drive off the enemy with gunfire, preferably in two units, the King advised. As of April 21, 1741 the regiment went its own way, though often with its sister regiment. The King did everything to increase the capability of the cavalry and always gave it a special mission to carry out during a battle instead of general participation. After spending the fall of 1741 waiting between Reichenbach and Neisse, it went to Olmütz in February of 1742 for the advance to southern Moravia. When the King moved into Bohemia early in April, it stayed behind in Prince Dietrich's Corps. During the march back from Fulnek to Troppau, four squadrons were attacked repeatedly by 3000 men in rugged country on April 19 and 20 in an attempt to detach it. In the successful breakthrough it lost 21 men and 59 horses. Afterward it stayed in Upper Silesia under Prince Leopold of Anhalt-Dessau. In 1744 it marched to Prague for the taking of the city, and survived the miserable withdrawal over the mountains on the border. At Hohenfriedeberg on June 4, 1745 it was in the second line on the left flank at Thomaswaldau facing a swampy terrain through which a passage to Teichau had to be found. Under Nassau and along with cuirassiers and its sister regiment, it made a victorious attack, lost only three officers and 37 men, and gained time for the infantry in the process. It marched through Landsberg/Warthe to join Dessau's Corps and was in Meissen on December 12. At Kesselsdorf it attacked the frozen slopes at a full gallop from the flank at the moment when the Saxon Grenadier Guards pushed the Anhalt Regiment back, cut them down and drove them out of the village. Under Kyau it encircled Kesselsberg from the east, leading to the capture of the town. Its commander, Colonel von Lüderitz, and six officers received the Pour-le-merite, and it could play the Grenadier March, but lost 51 dead.

4. Dragoner-Regiment
NON SOLI CEDIT
G. Dorn

After the march into Saxony on August 29, 1756 and the enclosure of Pirna, it broke forth without command in the second line of the left wing at Lobositz and threw the Austrians back anew. Its chief, General von Oertzen, its commander, General von Lüderitz, and 29 men fell. At Prague on May 6, 1757 it attacked on the left wing at Sterbohol after the cuirassiers were already in action and took several hundred prisoners, losing 59 men, until Zieten's encirclement became effective. At Kolin it was on the extreme left wing under Zieten in Normann's Dragoon Brigade, holding the superior enemy forces in check. Then it went with the King via Leitmeritz to the Saale. On September 17 it posed as infantry in Seydlitz's attack on Gotha, near Siebleben. At Rossbach on November 5 it was in the first line with its sister regiment and twice overrode the enemy, losing five officers. Back in Parchwitz on November 28, it fought at Leutzen on December 5 in the second line on Zieten's right wing, which put Nadasdy's Corps to flight. In 1758 it marched through Schweidnitz to the siege of Olmtz, then on August 11 from Landeshut to the Oder. At Zorndorf on August 25 it was with the Gardes du Corps and Gensdarmes under Lentulus, attacking the Russian cavalry and then the infantry in the Zabern Valley so successfully that the enemy wing was shattered. It captured seven flags and fourteen cannon, lost 89 men and received two Pour-le-merite and seventeen promotions. At Hochkirch it fought west of the town with the Zieten Hussars and Schönaich Cuirassiers, steadily beating back the Austrian grenadiers. On October 26 it took 500 prisoners at Ebersbach, near Görlitz and gained a Pour-le-merite. In 1759 it saw little action: Schmottseiffen, Sagan, Görlitz, Hoyerswerda and Strehla. In the attack on Kossdorf on Fenruary 20, 1760 its chief, General von Czettritz, a good friend of Seydlitz, was taken prisoner while carrying secret orders. After the cavalry battle at Göda on July 7 and the vain storming of Fresden, it was in the center at Liegnitz on August 15 but did not go into action. Via Hohenfriedeberg and Schweidnitz it moved to Wittenberg and Torgau, where it belonged to Holstein's reserve cavalry on the left flank. In 1761 it marched from the Elbe to Strehlen. Under Lentulus at Wahlstatt on August 15 it joined Dragoon Regiment 10 in a successful attack on Laudon's superior carabiniers and grenadiers. Then it was camped at Bunzelwitz until the Russians withdrew, afterward at Pilzen and Gross-Nossen. On August 16, 1762 it was led through Reichenbach at a gallop by the Duke of Württemberg to put the Archduke Joseph Dragoons to flight at Peilau under Lossow, with the loss of 200 men and three standards. It received a Pour-le-merite. Afterward it protected the siege of Schweidnitz until October 10. In 1763 it consisted of 683 Prussians, 20 Saxons and 255 'foreigners'.

In 1806 it separated on the return from Jena; two squadrons capitulated at Ratekau on November 7, three and the depot unit reached East Prussia and a reserve squadron was gathered in Danzig. In 1914 it was the Mounted Grenadier Regiment No. 3.

4. Dragoner-Regiment

DRAGOON REGIMENT 5

Commanders of the Regiment

1717	4/2	Colonel Achaz von der Schulenburg, later Lieutenant General
1731	8/7	Colonel Friedrich, hereditary Prince of Bayreuth, Margrave since 5/17/1735, later Lieutenant General
1763	May	Lieutenant General Friedrich Christian, Margrave of Bayreuth
1769	8/2	Lieutenant General Christian Friedrich Carl Alexander, Margrave of Ansbach and Bayreuth
1806	3/5	Major Luise, Queen of Prussia, position reserved for the queen by order of 8/4/1810

On April 2, 1717 the twelve cuirassier regiments, Dragoon Regiments 1/2 and 3/4 and the Wartensleben Mounted Regiment, which was disbanded in 1718, received orders to contribute non-commissioned officers and men for a new dragoon regimenton June 1, totaling 44 junior officers and 295 privates. They were to be "unmounted, without a uniform, with only an old uniform coat on their bodies, which people shall also be not more than forty and not less than twenty years old, be natives of the land, and shall also have done at least one campaign". Those were the orders for the formative cadre. On the next day the new chief, Colonel Achaz von der Schulenburg, was given the already chosen staff of officers. The 689 men needed to fill the ranks were gained through recruiting. The formation took place in Halberstadt, with four squadrons at first; in 1718 the fifth was added but at somewhat lesser strength. As of May 18, 1725 the regiment was expanded to ten squadrons of 110 privates each, not divided into companies, for which it gained eleven officers, fifteen junior officers, ten drummers and 450 privates. It gained 100 privates in 1726 and again in 1732. It was now divided into two battalions, one commanded by the regimental commander, the other by the next ranking staff officer. With the staff there were two majors and two adjutants. Its complement included 62 officers, 120 junior officers, ten flagsmiths, thirty drummers, eight medics, five oboists, 1320 privates and fifty supernumeraries, plus 1486 horses. Its Chief von der Schulenburg (1669-1731) had entered court as a Junker in 1688 after three years of study at Frankfurt/Oder and Wolfenbüttel, and become an ensign in 1690. He was a trusted friend of Friedrich Wilhelm I, who actively exchanged thoughts and opinions with him. As a Lieutenant General, he was appointed chairman of the court-martial of Crown Prince Friedrich II and Lieutenant von Katte, which declined to sentence the Crown Prince, sentenced Katte to life imprisonment and remained true to the King. As of 1733 its replacements came from the districts of Wollin, Daber, Lauenburg, Bütow and parts of Saatzig and Greifenhagen with the cities of Lauenburg, Bütow, Fiddichow, Pasewalk, Greifenhagen with the cities of Lauenburg, Bütow, Fiddichow, Pasewalk, Greifenhagen, Uckermünde, Gollnow, Gartz, Daber and Treptowan der Toll. Its garrisons were Halberstadt, Dornburg, Hamersleben and Veltheim, as of 1718 Kyritz, Lenzen, Königsberg/Neumark, Gartz, Arneburg and Ruppin, where it remained until 1724 minus Königsberg and Arneburg but plus Schwedt, Neustadt/Dosse and Penkun. As of 1733 it was in Pasewalk, Gartz, Gollnow, Uckermünde and Treptow an der Toll plus a few changing nearby towns. It went to Near Pomeraniain 1721 in exchange for the Crown Prince Mounted Regiment (No. 2). As of August 7, 1731 it bore the name of "Bayreuth"; on March 5, 1806 it was named "the Queen's Dragoons".

On December 16 it crossed the Silesian border on the left wing of the First Corps, enclosed Glogau and then, after being relieved by the Second Corps, marched to Breslau. At Mollwitz six squadrons fought on the left wing, which was not hit hard and was able to decide the battle; it lost 47 men and 80 horses. Four squadrons under General von Gessler did not arrive at the artillery park in Gohlau at the right time. Then it fought at Brieg, which surrendered on May 4. On May 17 two squadrons and Zieten's Hussars attacked a detachment under Baranyay at Rothschloss, west of Strehlau, and almost took them prisoner. After an engagement at Heinrichau, it left for northern Bohemia with Hereditary Prince Leopold. With ongoing instructions and bits of advice, the King kept the troops learning from their war experiences and adapting to the situation. In February of 1742 it moved to northern Moravia, but on May 13 it was back in Bohemia, at Chrudim. At Chotusitz on May 17 one battalion was on each flank, the right one in the second line, and so they joined in both attacks. In the first it met the Austrian infantry to the right of the Cirkwitz Pond, losing three standards when their bearers were killed, as was their commander, Colonel August Friedrich von Bismarck, the Chancellor's great-grandfather. On the left flank it met the enemy counterattack on the Brslenka. It lost fourteen officers plus many men and horses, more than the infantry. The King honored Captain de Chasot with the Pour-le-merite on the battlefield.

In 1744 it took part in the capture of Prague. On October 26, during the march back to the Elbe, it held off the enemy avant-garde at Kammerburg under Nassau, so that the meeting of the armies at Königgrätz was no longer prevented. Its finest hour struck at Hohenfriedeberg on June 4, 1745; on the left wing under Gessler, it had not made contact with the second line of cavalry and followed the second infantry line over the Striegau Water. So it found itself behind the gap at Thomaswaldau and under enemy fire. Since Dragoon Regiments 3 and 4 of the second line had joined with the cuirassiers, the enemy front was broken and there was no reason to wait, Colonel von Schwerin, its commander, and his officers decided to break into the enemy line between the Bredow and Münchow Brigades at a full gallop; the line broke, twenty battalions were smashed, 67 flags and five cannon captured and 2500 prisoners taken, with a loss of six officers and 88 men, 28 dead. In terms of timing, manner and success this attack was a masterpiece! The regiment won three Pour-le-merite and was praised by the King as "the Caesars of Hohenfriedeberg". Gessler became a count, Schwerin a general, other promotions resulted, and the trophies came in the form of a coat of arms and a seal. On September 23 it fought at Schatzlar and went with Lehwaldt's Corps through Bautzen and on to Meissen to meet the Old Dessauer on December 13. At Kesselsdorf it was on the inner left flank that scarcely got to fight before the Zschon Valley, where the terrain prevented the Saxons from being defeated.

At Lobositz on October 1, 1756 eight squadrons rode in the first attack under Kyau and helped the Gardes du Corps out of a bad situation, and two squadrons were in the second, against the

5. Dragoner-Regiment

King's will. It lost four officers, seven non-commissioned officers and 150 men. After that it was before Pirna until the surrender on October 16. During the Battle of Prague on May 6, 1757 it was in the cavalry corps of Prince Moritz of Anhalt-Dessau on the Sazawa, making a thoroughly unsussessful attempt to cut off the enemy's retreat to the south. After the siege of Prague was lifted on June 19, it marched back to Leitmeritz under Keith. It broke out of the Zittau area in August under the Duke of Bevern and reached Silesia. On September 7 it was in the Battle of Moys, near Görlitz, in which Winterfeldt fell. Via Liegnitz and Steinau, crossing the Oder twice, it reached a position on the Lohe west of Breslau. On the left wing at Kleinburg on November 22, it cut down a grenadier battalion, took hundreds of prisoners and captured four cannon. At Parchwitz on December 2 it joined the King's Army to fight at Leuthen three days later, on the outer left wing under Driesen, leading the final attack between Leuthen and Frobelwitz. It outflanked the Austrians and, along with C.R. 11, took two complete infantry regiments prisoner, cut down a battalion and captured four standards, nine flags and eight cannon. After the conquest of Schweidnitz in 1758 it went with the King to Olmütz, which it helped to enclose on the east bank of the March under its commander, Major General von Meier. On June 17 eight squadrons were attacked and lost 17 officers and 457 men, of whom twelve officers and 297 men were taken prisoner. Two squadrons were with Zieten at Domstadtl late in June and rescued part of the supply train. The regiment suffered long from its misfortunes. At Hochkirch it belonged to Retzow's Corps, which entered the battle at just the right time and prevented the worst. In the retreat it beat a cuirassier regiment and took 64 prisoners. In 1759 it did securing duty between the Bober and the Elbe. In 1760 it joined Prince Heinrich's Corps and was attacked at Kossdorf near Torgau on February 20, losing ten officers, 308 men and a standard. Its route led through Frankfurt/Oder, Landsberg, Meseritz, Glogau and Breslau to Hermannsdorf, where it joined the King on August 29. The Froidville Squadron joined Werner's advance on September 6 and tried for twelve days to relieve Kolberg, which had been surrounded by the Russians; leading to the enemy's flight and the saving of the city. Then it served at Schwedt, Schlawe and against the Swedes. As of October 7 the regimentwent with the King from Schweidnitz via Guben to the Elbe. At Torgau on November 3 it fought in Holstein's cavalry under Finckenstein; Colonel von Bülow and the First Battalion surrounded the Ahremberg Division on the enemy's right wing with four regiments, took whole battalions prisoner, captured ten flags and scattered the rest. The battle was almost won, but flank fire and counterattacks forced a withdrawal. The King was full of praise and gave eight Pour-le-merite. In 1761 it marched to Saxony with the King and was camped at Pilzen, Strehlau and Bunzelwitz without doing anything noteworthy. In 1762 it first went to Schweidnitz, the fate of which was determined by the Battle of Burkersdorf. In 1763 it consisted of 1616 Prussians, 50 Saxons and 255 'foreigners'.

In the campaign against the French Republic from 1792 to 1794 it repeatedly distinguished itself and received a total of twelve Pour-le-merite and twenty other medals. In 1806 it was with the main army at Auerstedt and fought its way through via Nordhausen and Halberstedt to Danzig with only 21 officers and 375 men. It remained in existence and was the Queen's Cuirassier Regiment No. 2 in 1914.

5. Dragoner-Regiment

DRAGOON REGIMENT 6

Commanders of the Regiment

1717	4/19	Major General Heinrich Jordan von Wuthenau, later Lieutenant General, died in a duel
1727	6/15	Colonel Hans Caspar von Cossell, later Lieutenant General
1734	9/23	Colonel Friedrich Christoph von Möllendorff, later Lieutenant General
1747	5/17	Major General Ludwig Wilhelm von Schorlemmer, later Lieutenant General
1760	11/9	Major General Carl Friedrich von Meier, later Lieutenant General
1777	1/24	Colonel Christoph Wilhelm Sigmund, Baron von Posadowsky, later Lieutenant General
1787	12/16	Major General Hans Ludwig von Rohr, later Lieutenant General
1790	4/9	Colonel Philipp August Wilhelm von Werther, later Lieutenant General
1803	4/30	Colonel Johann Kasimir von Auer, later Major General
1807	3/20	Colonel Christoph Johann Friedrich Otto von Zieten, later Lieutenant General

August II the Strong (1670-1733), Elector of Saxony since 1694, King of Poland since 1697, had to cut down his German regiments that were stationed in Poland in 1716. Since he was bound by contracts and did not want to lose them completely, he tried to rent them out to his neighbors for a number of years. King Friedrich Wilhelm I declared on May 22 of that year that he was ready to take four squadrons for a term of twenty years: "If I got such a regiment of dragoons with six hundred privates, Saxony would be a pleasure for me. For that I am willing to pay twenty Thaler per man. Their people can choose them, but no deserters; Poles, Russians, Saxons, Bohemians, Silesians and sons of my country, no other nation", he had already written on May 3. Probably valuable porcelain from the castles of Charlottenburg and Oranienburg was given in payment instead of money. For that reason they were later known in jest as "Porcelain Dragoons", or as Frederick the Great remarked, perhaps just because of their white and blue colors. But a complete regiment did not come; instead the seven cuirassier regiments each gave from 38 to 44 men, totaling 282, and the eight dragoon regiments each gave between 33 and 54 men, totaling 318 men. By proclamation of April 19, 1717 the new regiment was turned over to Major General Heinrich Jordan von Wuthenau, the soldiers were received in Baruth on May 1 of that year and formed into eight companies in Beelitz, Mittenwalde, Teltow and Treuenbrietzen on May 5. In September of 1718 it was expanded to five squadrons, in October of 1725 to ten of 110 men each, increased to 120 privates in 1726. After General von Wuthenau's death it was divided on June 15, 1727,just as Dragoon Regiments 1 and 3 were: One half stayed with the regiment with the kettledrums, five squadrons with the standards became Dragoon Regiment 7 under Colonel von Dockum. The regiment received Colonel Hans Caspar von Cossell as its chief. In 1733 and 1734 the squadrons were enlarged to 132 privates. By command of August 19, 1739 the division into ten companies was introduced again, to prepare for reforming into ten squadrons, which did not actually happen.

At the end of April 1734 it marched with the supply corps from Magdeburg to Heilbronn and joined the Imperial army during the War of the Polish Succession. Early in June Prince Eugene met the army there. Crown Prince Friedrich II also appeared on June 30, so as to learn military leadership from Prince Eugene. But he did not venture an attack; Philippsburg was lost. The Crown Prince gained only one impression of the Austrian army. On September 23 of that year the regiment was put under Colonel von Möllendorff, and as of October 1 it was strengthened by ninety men. Then it went to winter quarters in Westphalia. The 1735 campaign amounted to even less. At first it guarded the Rhine between the Main and Neckar, and then stayed at Leeheim, Heidelberg and Hockenheim, not far from Speyer, where the campaign ended. After being inspected by the King at Halberstadt, it marched to its garrison in East Prussia. At the end of 1740 it received five newly-established squadrons and then had a strength of ten.

Since 1733 its replacements came from the ever-threatened and sometimes surrendered East Prussia, so that in the Seven Years' War it had to be supplied with central German troops. Until 1806 the canton included the districts of Barten, Gerdauen, Grünhoff, Fischhausen, Georgenburg, Insterburg and Soldau with the cities of Fischhausen, Pillau, Labiau, Allenburg, Goldap, Marggrabowa (later Treuburg) and parts of Königsberg. From 1717 to 1730 it was garrisoned in Insterburg, Heiligenbeil, Pillkallen and Stallupönen, from 1731 to 1733 in the same cities minus Heiligenbeil. Until 1743 it was in Insterburg, Gumbinnen and Darkehmen. From 1746 to 1806 it was in Königsberg, Wehlau, Allenburg, Labiau, Drengfurth and Gerdauen, without Drengfurth from 1788 to 1793 and with Darkehmen as of 1800. As of November 1, 1721 the King had two companies of hussars set up in the regiment, the very first in Prussia, called "Wuthenau's Hussars"; they were stationed in Memel and Tilsit and, when the regiment was divided in 1727, went along to Dragoon Regiment 7 for five years and then returned. At the end of 1735 the squadrons, now numbering three, were again put under the chief of Dragoon Regiment 7 as the "Prussian Hussar Corps".

At the beginning of the war it was under Buddenbrock in East Prussia. Early in June of 1741 it moved from there, along with Hussar Regiments 1 and 3, to provide reinforcements in Silesia, where the first battles had been fought. The summer brought only minor skirmishes of light troops. At the end of December two squadrons went to Moravia in Schwerin's Corps to occupy Olmütz. In February of 1742 the other eight squadrons followed them in the King's Corps for the advance into southern Moravia. The winter campaign led to the surrounding of Brno. In the latter half of March the King gave the cavalry numerous instructions to prepare them for the coming battles. On April 1 he wrote to Hereditary Prince Leopold: "I am better satisfied with our cavalry officers than last year". When the King moved into Bohemia, Prince Dietrich had to evacuate Moravia step by step. On April 10 the regiment fought at Austerlitz under its chief, Major General von Möllendorff, smashing light enemy troops in efficient fighting order. At the end of April it took up residence in the Jägerndorf-Troppau area, for it needed to revive. While the King assembled his troops for Chotusitz, it secured Upper Silesia, as of early May under Prince Leopold of Anhalt-Dessau. The enemy carried on a small-scale war here because its main forces were needed against the French. The King reminded the regimental commanders "that their own work should be devoted to making capable cavalrymen out of the common soldiers, as adroit as the hussars". In August of 1744it moved from East Prussia to the Mark Brandenburg because of Saxony's activities; as of September 25 it was under Prince Leopold. So it was spared the bitter withdrawal from Bohemia in the fall. In April of 1745 it, instead of C.R. 3 and 6, went to Silesia to Steward Waldburg's Corps. On May 22, under Winterfeldt south of Landeshut, it turned the Nadasdy Corps' attack into a victory. Led by Major General von Stille in the second battle line, it

6. Dragoner-Regiment

chased Hungarians and Croats over the Reichhennersdorf Mountains as far as the Abbey of Grssau. At Hohenfriedeberg it was on the right wing in Posadowsky's second line, against which the Saxon cavalry's attack broke. In the tumult of the counterattack it and C.R. 1 captured four standards and kettledrums each. The enemy fled.

In 1756 it remained in Field Marshal von Lehwaldt's Corps in East Prussia. It became ever clearer that the Russians would intervene. At Gross-Jägersdorf on August 30, 1757, on the left wing under Lieutenant General von Schorlemmer, its chief, it broke through the Russian cavalry between the Norkitt Woods and the Pregel, pushed deep into their position and captured a battery. In the forest battle it could no longer attack; at the end it could only cover the infantry's retreat. It lost four officers, 105 men and 164 horses. After the Russians had left East Prussia, the King ordered the withdrawal to Pomerania. Until February 1758 the Russians occupied the entire province. In 1758 it was in the Pomeranian Corps of Lieutenant General the Count zu Dohna, who left Near Pomerania and retreated to Frankfurt/Oder. On August 21 the King arrived from Silesia; he underestimated the Russians. At Zorndorf on August 25 it served as the reserve behind Kanitz's left flank, which followed Seydlitz's attack as pursuit. After that it supported the infantry, which had been shaken by Demiku's attack. The losses amounted to fourteen officers, 223 men and 423 horses, including Major General de Froideville, its commander. At the end of October it marched to the Elbe and attacked the enemy in the flank across a ford at Eilenburg. In 1759 it experienced the failed advance on Posen and the hopeless attack on the Russians on the Palzig Heights at Kay on July 23.On the left flank under Seydlitz at Kunersdorf, it threw itself against the enemy attack but could not straighten out in the Kuhgrund. Lieutenant General von Platen led it against the great battery on the Spitz-Berg, until the mass attack from the left drove the Prussian cavalry from the field. Eighteen officers and 234 men fell. At Meissen on September 21 it successfully fought off the attack of Hadik's superior forces and, in a raid under Kleist, destroyed the great magazine in Aussig. In 1760 it was in the King's Army behind the Triebisch, then before Dresden, the taking of which failed in July. When the King went to Silesia, it stayed at Schlettau, near Meissen, under Hülsen. On August 20 at Strehla, the battalion under Major von Marschall cut down the Esterhazy Regiment and eight companies of grenadiers, drove two cavalry regiments to flight, and captured the Zweibrücken Dragoon Regiment with two flags and a cannon. By so doing it saved the Hülsen Corps, though it lost four officers and ninety men. Marschall received the pour-le-merite. The march back via Wittenberg, before fourfold superior troops, led to Beelitz. At the limit of its usefulness in 1761, it saw defensive action on the Mulde with the Saxon Corps. In 1762 it had 768 vacancies! It had bled dry. On May 12 it crossed the Mulde at Döben—Major von Treskow received the Pour-le-merite—and then went to the Pretzschendorf camp in narrow column formation for defensive reasons, and on to Freiberg. On October 29 Major Marschall von Biberstein led the Second Battalion in the Forcade Column in an attack on the heights of Gross-Schirma.

In 1806 it fought in L'Estocq's Corps at Soldau, in 1807 at Preussisch Eylau and Heilsberg, where the First Battalion smashed two French cuirassier regiments with many losses. As of 1845 Field Marshal the Count von Wrangel was its chief. In 1808 the First and Second Battalions formed Cuirassier Regiments 3 and 4.

6. Dragoner-Regiment

DRAGOON REGIMENT 7

Commanders of the Regiment

1727	6/15	Colonel Martin Arend von Dockum, later Major General
1732	7/11	Colonel Eugen, Prince of Anhalt-Dessau, later of I.R. 8
1737	12/23	Colonel Christoph Friedrich von Thümen, later Major General, later of Garrison Regiment 5
1741	8/19	Colonel Ernst Friedrich von Werdeck, later Major General, died 5/17/1742 at Chotusitz
1742	5/19	Colonel Friedrich Alexander von Roell, later Lieutenant General, died 12/13/1745 at Meissen
1745	12/26	Major General Erdmann Ernst von Ruiz, later Lieutenant General
1756	8/3	Major General Christoph Friedrich Stephan von Plettenberg, later Lieutenant General
1761-	1763	vacant
1763	9/10	Colonel Levin Friedrich Gideon von Apenburg, later Major General, later of C.R. 1
1781	3/19	Colonel Gottlieb Matthäus Siegfried von Borck, later Major General
1790	9/25	Major General Johann Ernst von Zabeltitz
1792	8/31	Major General Wilhelm Friedrich von Schenck, later Lieutenant General
1803	4/30	Major Generall Friedrich Ludwig von Passau
1805	10/3	Colonel Adrian Gottlieb von Rhein
1806	11/12	Colonel Joseph Theodor Sigmund von Baczko, later Major General

The regiment developed from Dragoon Regiment 6 taken over from Saxony and reorganized under Major General von Wuthenow, which was formed with eight companies on May 5 of that year in Beelitz, Mittenwalde, Teltow and Treuenbrietzen. Then it was garrisoned in East Prussia. In September of 1718 it was increased to five squadrons, in October of 1725 to ten, and on June 15, 1727 it was divided. Five squadrons formed the new regiment under Colonel Martin Arend von Dockum, and Tilsit became its garrison. After the death of August the Strong it was transferred to the Prussian Supply Corps along with Dragoon Regiments 2 and 6 . In January of 1734 it left there, gathered at Halberstadt until April, and met the Imperial army led by Prince Eugene of Savoy in Heilbron nearly in June. The expectation of great success was not fulfilled. According to orders of August 19, 1739 it was divided into ten companies and supplied with officers to prepare for division, having 31 officers, sixty non-commissioned officers, five flagsmiths, twenty drummers, five medics, 660 privates and 30 supernumeraries, 811 men in all, and 824 counting the staff. From 1727 to 1732 the Tilsit "Prussian Hussars" belonged to the regiment, likewise from 1735 to 1737. As of 1733 its replacements came from the districts of Tilsit and Memel and parts of the district of Insterburg, with the cities of Tilsit and Memel. Its garrisons were Tilsit from 1727 to 1733, Köpenick in 1734, Melle, Quakenbrück, Bramsche and Badbergen in 1735, Tilsit from 1736 to 1740, Berlin and Potsdam in 1741, Berlin, Köpenick, Treuenbrietzen, Beelitz, Mittenwalde and Fürstenwalde in 1742, Tilsit in 1743, Magdeburg in 1744, and Tilsit from 1746 to 1806. Its chief from 1742 to 1745, Friedrich Alexander von Roell (1676-1745), brave and perspicacious, had risen from the ranks and was later ennobled.

Like Dragoon Regiment 6, it received the command to set up five new squadrons at the end of 1740, for which recruiters from the old regiments went out into the Empire. In January and February of 1741 its reorganization was completed, so that it was able to go to East Prussia with ten squadrons under its chief, Major General von Thümen, and guard against Polish raiders. That June it marched to the Mark to join Prince Leopold of Anhalt-Dessau in camp at Gttin. Here Colonel von Werdeck took it over. On March 13, 1742 the King ordered this corps to the Jägerndorf area by the end of April, so that it could advance east of the March. But in April the King moved his center of operations to Bohemia. The regiment went with him to Chotusitz. On May 17, on the left flank behind the cuirassiers, it attacked north of the village, but was driven back by a counter-attack and separated from the cuirassiers. Grenadiers tore the guard standard from the badly wounded Junker von Roop. The regiment threw von Birkenfeld's Cuirassier Regiment back, but lost 156 dead, 77 wounded and 283 taken prisoner, 516 soldiers in all, including its chief, Major General von Werdeck, as well as 156 horses, one-fifth of all the cavalry losses. Two Pour-le-merite were its thanks. In April of 1744 the King wanted to double the thirty dragoon squadrons in East Prussia. Except for the formation of Dragoon Regiments 9 and 10, though, it only amounted to the formation of an independent Dragoon Regiment 8 out of the second, third, fifth, eight and tenth squadrons on November 1, under the former regimental commander, Colonel von Stosch. Meanwhile it had been transferred from East Prussia to the Mark Brandenburg, where Prince Leopold of Anhalt-Dessau had taken command again on September 25, because of the Saxon situation. So it was spared the Bohemian campaign. As of late April of 1745 the corps gathered again at Magdeburg. After getting ready to march as of July 27, it left on August 31 to go to camp at Dieskau, and went on to Halle on November 23. When on December 13 it was passing through a ravine along the Elbe on its way from Zehren to Meissen at the end of a long column of cavalry, it was attacked by the Rutowsky Dragoon Regiment of Saxony and suffered losses; men were taken prisoner and two standards were captured. Its chief, Lieutenant General von Roell, in his wagon because of ill health, was shot. At Kesselsdorf on December 15 it fought on the right wing under Kyau, surrounding the village from the east, disposed of the Carabinier Guards and the Mounted Grenadiers, rode down the Foot Guard and the Niesemeuschel Regiment, and captured a guard flag, a standard and kettledrums. The focal point of the enemy position was taken.

In 1756 it was with Lehwaldt's Prussian Corps, which opposed the Russians as of July 1757. At Gross-Jägersdorf on July 31, fighting on the right wing under the Duke of Holstein, it shattered three cavalry regiments and pursued them to behind the infantry. Major von Korff took a ten-gun battery and smashed a battalion with two squadrons. The other three squadrons joined Dragoon Regiment 8 in attacking the Wologda and Susdal Infantry Regiments and causing them heavy losses. Korff was promoted preferentially; the losses added up to 136 men. As of October 1 it marched into Pomerania and to Demmin. In 1758 it stayed in the Pomeranian Corps, now under Dohna, who evaluated the Russians realistically. When they appeared at Meseritz, Dohna protected Frankfurt/Oder. Four days after the King's arrival at Küstrin on August 21, it joined Dragoon Regiments 6 and 8 to form the reserve behind Kanitz's left wing and attack with Seydlitz, struck several Russian regiments and took five cannon, losing 63 men. The enemy wing was beaten. Next it knocked the infantry out. Then on September 22 it drove the Swedes back at Zehdenick, took three officers and 300 men of the Swedish Bodyguard Cuirassiers prisoner at Linum, liberated Fehrbellin on the night of September 28 and, under Wedell, drove the Swedes back to Prenzlau until October 17. Then it had to leave for the Elbe to protect Torgau on November 12. At Eilenburg it captured two cannon. Ordered back to Pomerania, it stormed Damgarten under Kleist on January 1, 1759, and later Anklam, to safeguard the Peene line. Back in Saxony at the end of August, it took part in the night

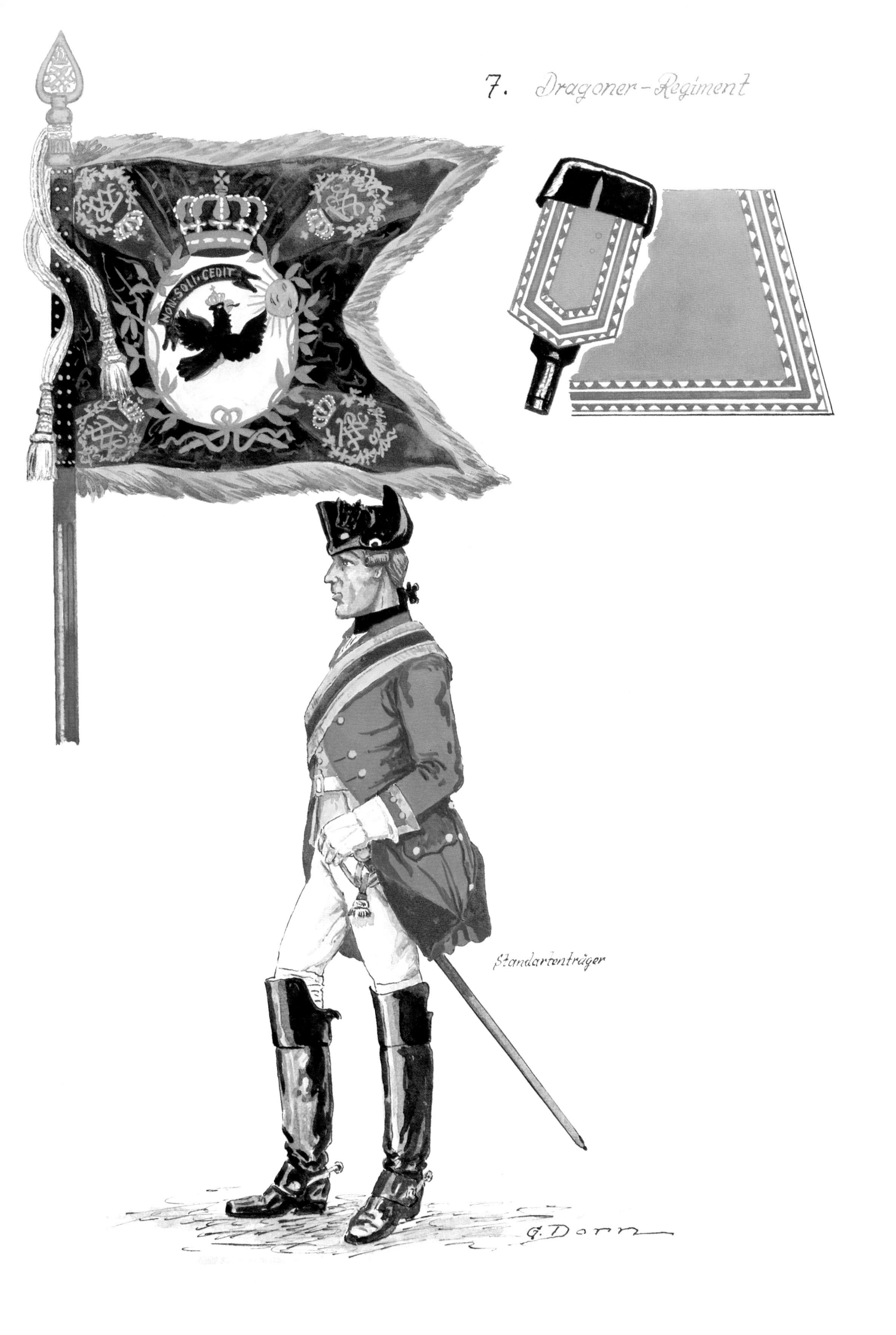
7. Dragoner-Regiment
NON·SOLI·CEDIT
Standartenträger
G. Dorn

storming of Torgau on August 29, and at Grossenhain on September 4 it shattered the Palatine Hussar Regiment, losing 368 men and 500 horses. At Torgau four days later it circled the enemy to the north and attacked from the rear with much noise, capturing eight guns, sixteen ammunition wagons, 26 officers and 850 men while losing four officers and 186 men. Torgau was saved. In battle at Korbitz on September 21 it pushed the Serbelloni Cuirassier Regiment into a narrow pass, causing heavy losses and taking ten officers and 64 men prisoner, while losing eight officers, 180 men and 68 horses.

In 1760 it belonged to Stutterheim's Corps which drove the Swedes back from the Peene to the Ucker, beginning in mid-August. With the Belling Hussars, into whose hands Blücher fell here, it led the skirmish at Kabel Pass near Friedland on August 27. Then it had to return to Templin slowly. When the Russians under Totleben appeared before Berlin on October 3, the corps—now under Duke Friedrich Eugen of Württemberg—hurried to the scene and defended the city until October 9. The regiment lost two officers and eighty men. After the arrival of an Austrian corps, it left Berlin and marched via Belzig to Kemberg, led the advance to Leipzig under Linden against the Imperial army, and stayed in Eilenburg to defend the move back during the Battle of Torgau. In 1761 it was in the Pomeranian Corps at Belgard and Körlin, south of the fortified camp before Kolberg, until mid-August. When the Russians attacked early in September, the cavalry under Werner broke out to disturb the enemy from the rear. At Garrin on September 6 it struck down the Archangel gorod Dragoon Regiment and took three standards. When Werner was captured at Treptow, Dragoon Kleibitz captured Colonel Count Wittgenstein. It moved on forward with Platen to Körlin, where it captured two guns. On October 2 at Spie it broke through into surrounded Kolberg, but had to break out again with Platen. It reached Gollnow on October 22, just in time. After the Russians withdrew, it moved ahead to Greifenberg on November 14. The breakthrough at Spie on December 12 failed; Kolberg could not be saved. It had lost one officer, 136 men and 154 horses. In 1762 it stayed at Malchin until January 10, then marched to Saxony. On May 12 it broke through across the Mulde and stormed the Klingenberg battlements that had been lost by the Bähr Grenadier Battalion, taking 500 prisoners. From October 14 to 16 it fought at Brand; the Mulde could no longer be held. At Freiberg on October 29 it attacked St. Michael, near Brand, at night in Kleist's first column. When the attack came to a stop, it encircled the enemy from the south, threw back the Bayreuth Cuirassiers, smashed the Salm Regiment and captured eight cannon, while two regiments overrode Infantry Regiments 33 and 51, taking 17 officers and 700 men prisoner. It received two Pour-le-merite and lost one officer, 66 men and 72 horses.

In 1763 the regiment consisted of 688 Prussians, 17 Saxons and 162 'foreigners'. In the Seven Years' War it lost twenty officers, 87 non-commissioned officers, 1652 men, 1989 horses, but only six deserters in Pomerania in 1761. It destroyed five cavalry and eight infantry regiments and took 54 officers, approximately 4000 men, 1000 horses and 33 cannon, very noteworthy for its deployment in minor action, and recognized by the King. In L'Estocq's reserve corps in 1806, it fought at Thorn and Soldau, in 1807 at Preussisch Eylau and Heilsberg. Carried over in 1808, it was Dragoon Regiment in 1914.

7. Dragoner-Regiment

DRAGOON REGIMENT 8

Commanders of the Regiment

1744	11/1	Major General Friedrich von Stosch
1751	12/24	Major General Adolf Friedrich von Langermann
1757	3/4	Major General Dubislav Friedrich von Platen ("Old Platen"), later General of the Cavalry
1787	7/6	Major General Carl Wilhelm von Brausen
1790	9/25	Major General Georg Friedrich von Bardeleben, later Lieutenant General
1801	4/26	Lieutenant General Carl Gottlieb von Busch
1803	5/12	Colonel Christian Carl von Esebeck, later Major General

As Dragoon Regiment 7 was formed in 1727 out of Dragoon Regiment 6, which was established in 1717, so this regiment derived in turn from Dragoon Regiment 7 by division in 1744, a proliferation of dragoons in two generations, so to speak. The rebuilding of the dragoon forces had been a basic necessity since 1715-1716, the founding of this regiment had been prepared for practically with the planned expansion of the other regiments to ten squadrons in 1739-1740, but in 1744 it had been limited specifically to East Prussia. The first use of the parent regiment 7 in the War of the Polish Succession as part of the Imperial army on the Rhine in 1734-1735 brought it nothing but marching experience and acomparison to the other contingents. It got its baptism of fire in the Silesian Wars only in mid-1742 at Chotusitz, shortly before the first war ended, losing over 500 men, many more captured than killed, even though it was praised by the King. In the army expansion planned by the King for April of 1744 it was intended—after the establishment of Dragoon Regiments 9 and 10, each supplied with ten squadrons by December 1, 1743—to expand each battalion of Dragoon Regiments 6 and 7 into a regiment of ten squadrons in order to double the number of East Prussian dragoons. But when the Second Silesian War began that summer, these plans were at first cut back in part and then completely abandoned after the unfortunate and costly course of the Bohemian campaign in 1744, because all strength was needed now to rebuild the army and all attainable means were needed to get through the coming 'war year of 1745 successfully, no matter what. On September 12 of that year the King still ordered the funds for equipment for the doubling of Regiments 6 and 7, and had the money ready for their support as of October 1. In dividing the prisoners from Prague, which took place on September 16, 1744, the King declared: "Then 1700 men will be chosen to be dragoons for the new regiments; of them Möllendorff (D.R. 6) gets 850 and Röhl (D.R. 7) 850". Thuson November 1, 1744 Colonel Friedrich Stosch, regimental commander since 1734 as a lieutenant colonel, received squadrons 2, 3, 5, 8 and 10 as the new Dragoon Regiment 8 and remained its first chief until he retired as a major general in 1752. The original intentions of an increase were not taken up later either. The formation took place in Magdeburg. A new establishment in a strict sense was not necessary, for the men knew each other, the ranks were filled, and the regiment was ready for action. Its chief from 1757 to 1787, General Dubislav Friedrich von Platen, ended his career as General of the Cavalry and Governor of East Prussia in difficult times. Its replacements came from parts of the district of Insterburg with the cities of Insterburg, Pillkallen, Ragnit, Schirwindt and Stallupönen from 1744 to 1806. Its garrison was always Insterburg, and also Ragnit from 1746 to 1755.

At the end of April 1745 it went—along with Dragoon Regiments 7, 9 and 10—to join the corps of Prince Leopold of Anhalt-Dessau at Magdeburg; the corps consisted of twelve musketeer and four grenadier battalions and thirty cavalry squadrons and was meant to prevent Saxony from openly entering the war after Saxon troops had turned up in Bohemia as "Helpers of the Queen of Hungary" In case the feared advance out of Upper Lusatia—at that time the Görlitz-Bautzen area—came toward Lower Silesia, the Dessauer was to "treat the Saxon lands, as far as possible, with hostility". Most of the regiments had already been stationed in the Mark. On August 26 the corps received its first strengthening at Köthen, on October 6 its second at the Dieskau camp near Halle, on December 13 its third at Meissen. It was now about 30,000 men strong. The King urged action; his financial situation was bad, Russia's attitude uncertain, a successful conclusion urgent. Since the Prince hesitated, he appeared with his army at Königsbrück to guard the rear. At Kesselsdorf on December 15 it experienced its first battle as a regiment. It was on the inner left flank in the first battle line, at the impassable Zschon Brook before Zöllmen, which only the infantry laboriously conquered in close combat under Prince Moritz of Anhalt. If the cavalry that was here under Lieutenant Generals von Rochow and Wreech had gained the icy ground at the right time, the Saxon army would have been thoroughly smashed, Jany judged. In comparison to the infantry, their losses were few.

Years of practical war experience, practiced maneuvers, instructions from the King and his constant instructional effect, plus a series of outstanding cavalry leaders, had created a spirited and energetic utilization of men and horses that made the Prussian cavalry the equal of the Austrian, and even superior. In his instructions of December 14, 1754 for the next spring's training, the King advised the regiments by way of preparing for the battlefield: "In winter the men must have ridden individually enough so that they can immediately begin as a unit in the spring. As soon as the weather improves, the squadrons are to be drilled as units . . . The commanders will soon become aware of which people must be taken in hand individually, whereby whatever is lacking must be corrected. It will also be clear which horses can no longer go on but must be sent to the rearmost units. Furloughed men and supernumeraries must become masters of their horses". Then came squadron and regimental drills. "When the attacks' fluctuations, marching forward and back, digging trenches and everything else goes well in the regiment, maneuvers can be made such as foraging, crossing bridges, also marching with the regiment". Whenever they marched out, the regiments were to have advance and rear guards and flank security.

In 1756, along with Dragoon Regiments 6, 7, 9 and 10, it was in Field Marshal von Lehwaldt's Prussian Corps in East Prussia, waiting for the Russians, who took Memel only on July 5, 1757. As of June 6 it camped at Insterburg, moving to Wehlau in mid-August. At Gross-Jägersdorf on August 30, it and Dragoon Regiment 7, together on the right flank under the Duke of Holstein, first attacked Sybilski's cavalry, then the Wologda and Susdal Infantry Regiments, and beat them down for the most

8. Dragoner-Regiment
PRO·GLORIA·ET·PATRIA
G. Donn

part. Its losses amounted to only 33 men and 27 horses. The long and bloody forest battle, the increasing effect of the enemy's superior forces, and panic led to a withdrawal. Platen received the Pour-le-merite. Nevertheless, the Russians evacuated East Prussia. Lehwaldt followed them as far as Tilsit. As of October 6 Lehwaldt turned toward Pomerania to take on the Swedes, who had meanwhile occupied Wollin, Demmin and Anklam. When the corps arrived, they evacuated the Oder islands and the Peene line as far as Stralsund. In 1758 Dohna took command of the corps, which was to fight off the Swedes first and then turn against the Russians if they advanced into the Neumark or Silesia. When they appeared on the Warthe, Dohna moved back to Frankfurt/Oder via Eberswalde. On August 22 the King joined him at Küstrin. At Zorndorf the regiment was in reserve behind the left wing, then was brought up from Zorndorf by Prince Moritz of Anhalt to join the great attack under Seydlitz and plunge into the infantry "uncommonly well" and with the greatest bravery. Struck in the flank and front, the whole enemy flank was shattered. It lost nine officers, 119 men and 229 horses. When the Russians besieged Kolberg, it drove them away at Greifenberg under Platen on October 26 and relieved the surrounded city. The Russians went back to the Vistula.

In 1759 it secured the Peene line between Anklam and Demmin under Kleist. On January 16-17 Lieutenant von Manstein crossed the Peene with twenty men and took an advanced battlement in hand-to-hand combat, leading to the capitulation of Demmin. On July 23, now under Wedell in Dohna's army at Kay northeast of Züllichau, it repeatedly attacked the more than doubly superior Russian forces on the fortified Palzig Heights from the front and over the Zauche lowlands, but ran into the counterattack. At Kunersdorf it was on the left flank in Finck's Corps, which went over the Trettin Heights from the north. The attack came to a stop on the Kuhgrund and was thrown back by a vigorous attack of the enemy cavalry. In 1760 it was sent to Fouqué's Corps, which was to defend Silesia at Löwenberg and Landeshut with 14,500 men. One squadron was sent to Neisse, four were at Landeshut on June 23 when Loudon attacked with threefold superiority. It lost 255 men but fought its way through to Zieten along with Hussar Regiment 6, losing two standards and the kettledrums. Captain de Thoyras did an exemplary job. At Neumarkt on August 18 it and Hussar Regiment 6 drove back two Austrian dragoon regiments, then it fought at Zobten and, in September, at Dittmansdorf southwest of Hohgiersdorf. Its last three squadrons went with two squadrons of Cuirassier Regiment 9 to guard Silesia in Goltz's Corps when the King struck Torgau. In 1761 three of its squadrons belonged to the King's army but took part in Platen's advance into Far Pomerania, taking away the Russians' gigantic supply train at Gostyn on September 15. On October 2 it broke through to Kolberg and soon afterward to the southwest, a luckless campaign. As of March 6, 1762 there were again recruits and remounts from East Prussia. The regiment saw action in Silesia at Warmbrunn, Burkersdorf, Reichenbach and Schweidnitz.

In 1772 the King judged very harshly: "At no time have I had reason to be satisfied with your regiment, because it always ran off and let the Cossacks chase it away. These conditions must inspire a keen displeasure in me". In 1806 it fought in L'Estocq's reserve corps at Thorn and Soldau, in 1807 at Braunsberg and Königsberg. Cuirassiers since 1819, it was Regiment No. 5 in 1914.

8. Dragoner-Regiment

DRAGOON REGIMENT 9

Commanders of the Regiment

1741	4/16	Lieutenant General Hans Friedrich von Platen
1743	10/27	Colonel Georg Ludwig, Duke of Holstein-Gottorp,later Lieutenant General
1761	4/9	Colonel Nikolaus Alexander von Pomeiske, later Lieutenant General
1785	8/23	Colonel Johann Bogislav von Zitzewitz, later Major General
1788	12/1	Major General Hans Friedrich Heinrich von Borstell, later of C.R. 7
1792	11/12	Colonel Hieronymus von Brückner, later Major General
1798	3/1	Colonel Friedrich Wilhelm, Count von Herzberg, later Lieutenant General

The regiment originated—as did Dragoon Regiment 10—from Dragoon Regiment 1 of Ansbach, founded in 1689 and enlarged to five squadrons in 1718. After Dragoon Regiment 8, it and its sister regiment are the second youngest dragoon regiments. On September 1, 1722 a squadron of light dragoons, numbering 150 privates, was added to Dragoon Regiment 1. The description "light" referred to the size of the dragoons and their fast horses; they were intended less for battle use than for reconnaissance and mobility. According to the regulation of 1727, a lieutenant with twenty to thirty light dragoons was to ride ahead of every column when the army was on the march, open the path with tools, cut through hedges and set up light bridges. Such a unit was also to follow at the end of every column to collect and bring back sick or slow men. Over the years these original purposes diminished more and more. When Dragoon Regiment 2 was separated in 1725, it stayed with the regiment and was enlarged in 1729 to two squadrons of 130 privates each. In 1734 the two squadrons were expanded to five, so that the regiment had five heavy and five light squadrons, whose officers were ranked separately. According to the plans of 1739, Dragoon Regiment 1 not only received five newly established squadrons in 1740, but its five light squadrons were also doubled in Pomerania, so that for a time it consisted of five heavy and ten light squadrons. By the end of 1740 the heavy squadrons were separated from the light ones and put into the Second Corps in Silesia. The ten light squadrons, on the other hand, left East Prussia early in April of 1741 and went to the Göttin camp to join the Observation Corps of Prince Leopold of Anhalt-Dessau, which was transferred to Grüningen on September 12 of the same year and disbanded at the end of October without seeing action. The time spent in camp was devoted to intensive drill and training. Since April 16 of that year it was separated from Dragoon Regiment 1 under its chief since 1725, General von Platen, and made independent, while Dragoon Regiment 1 was put under Colonel von Posadowsky. Its final disposition, though, had not yet been determined. In the first half of the war year of 1742 it was partly on the lower Oder. In the course of expanding the dragoons in East Prussia, Platen's Regiment was divided into Regiments 9 and 10 on December 1, 1743. Since Dragoon Regiments 11 and 12 already existed, the series was finished. Dragoon Regiment 9 received Colonel Georg Ludwig. Duke of Holstein-Gottorp, as its chief for eighteen years; he was a close relative of the Russian royal couple Peter III and Catherine II, an Imperial Russian field marshal, father of the second Duke of Oldenburg, and died on September 7, 1763. The regiment's replacements came from the districts of Bütow, Lauenburg and Elbing and parts of Rummelsburg from 1741 to 1743, and from the districts of Mohrungen and Marienwerder with the cities of Marienwerder, Riesenburg, Bischofswerder, Freystadt, Rosenberg, Garnsee and Liebemühl in West Prussia from 1744 to 1806. Its garrisons were Angerburg and Insterburg in 1741, Wriezen, Angermünde, Pasewalk, Schwedt, Freienwalde and Gartz in 1742, Riesenburg, Marienwerder, Freystadt and Deutsch Eylau in 1743, Magdeburg in 1744, then Riesenburg, Liebemühl, Freystadt, Deutsch Eylau and Marienwerder from 1746 to 1790, though with Bischofswerder instead of Freystadt as of 1780 and Christburg in place of Marienwerder as of 1789; they remained unchanged until 1806 except that Saalfeld was substituted for Liebemühl as of 1802.

The regulation made in 1743, amplified in the same year and amended in 1744 brought about a basic change in the double role of both infantry and cavalry use and led to the cavalry being used strictly as such in terms of both training and field service. In 1744 it left East Prussia along with Dragoon Regiments 6, 7, 8 and 10, went to the Mark Brandenburg as a precautionary move, and thus stayed within that province. East Prussia was cut down to only the garrison regiments. When the indefinite stance of Saxony turned to hostility, the King once again gave command to Prince Leopold of Anhalt-Dessau on September 25. In 1745 the same troops—this time minus Dragoon Regiment 6—gathered at Magdeburg at the end of April to keep Saxony out of the area if possible. By possessing Lusatia it dominated the broadest access to Silesia. After the King's victories at Hohenfriedeberg and Soor, it was at Halle on November 23. Despite three rounds of reinforcements, Prince Leopold remained cautious at first. Haste was necessary, though; the King was impatient. On December 9 he wrote: "And is my field marshal the only one who cannot or will not understand my orders?" When the regiment marched along the Elbe toward Meissen at the rear of the cavalry column through the ravine at Zehren on December 13, it and Dragoon Regiment 7 were attacked by the Saxon Lieutenant General Baron Sybilsky with the Rutowsky Light Dragoon Regiment and two uhlan units, losing dead, wounded, prisoners, two standards and its silver kettledrums. At Kesselsdorf it was on the outer right wing in the second battle line under Kyau and attacked the enemy infantry wing, dependent on the village, in the flank and rear. In peacetime it belonged to the Prussian Corps under Field Marshal von Lehwaldt as its commanding general, staying together in 1756.

In 1756 the corps was strengthened to 29 battalions and sixty squadrons with about 30,000 men. Other than furloughed men and doubled supernumeraries, whom the regiment retained, it received, like all other regiments, 100 to 150 recruits from its canton, who were trained and outfitted in Königsberg. They were then made ready to ride with horses from the Tilsit area. Lehwaldt had full power for the operation: "Your corps is weak, to be sure—but I still hope that you will soon have it ready, and particularly that your cavalry has the opportunity to defeat the enemy". This came only at the end of June 1757, when the Russians appeared on the East Prussian border. Lehwaldt moved against them south of the Pregel at Wehlau only at the end of August. At Gross Jägersdorf on August 30 it broke through all three Russian battle lines on the left wing under

9. Dragoner-Regiment

Schorlemmer and captured eight cannon, which it could not bring back because it lacked teams of horses. Its losses amounted to five dead, fifty wounded and thirty horses. Despite the cavalry's success, the bravely fighting infantry was denied a victory. The corps followed the withdrawing Russians to Tilsit, but turned on October 6 to move against the Swedes in Pomerania. At the beginning of 1758 it was before Stralsund; it did not cross to Rügen. When Duke Ferdinand of Braunschweig and the "Allied Army" relieved the King's west flank against the French, he had it and Dragoon Regiment 10 go with him via Artlenburg, under Duke Georg Ludwig for reinforcement, since light cavalry was lacking. After contact with the enemy at Nienburg, its advance led to the Rhine via Minden and to Goch on June 3. At Krefeld on June 23, with Dragoon Regiment 10 and followed by the Hessian Bodyguard and Dragoon Guard Regiments, it attacked the French infantry from the rear, drove them out of their stubbornly defended battlements after a bloody fight, and captured a standard and the drums of Royal Roussillon Cuirassier Regiment, with only modest losses. Because superior French troops moved into Hesse, it had to recross the Rhine in July and cover the retreat, as it did at Borck on September 28 and 29. In 1759 it was at Freiensteinau in North Hesse. When Duke Ferdinand marched toward Frankfurt that spring to strike the French army singly, he met the French, 30,000 men strong and under the Duke of Broglie, at Bergen-Enkheim in such a superior position that they were able to drive off three attacks and take their offensive to the Weser. The regiment lost thirteen men and 22 horses. After the French had captured the fortress of Minden on July 8, they were brought into battle on August 1. The regiment took the La Marine Brigade prisoner and captured two flags and ten (of thirty) cannon, losing 67 men and 115 horses. The Duke awarded cash to the regiment; the King granted them the Cuirassier March for Krefeld.

After his losses, the King called it back to his army in 1760 despite the Duke's request when he moved from Landeshut to Saxony in June. Here it took part in the unsuccessful attempt to take Dresden and turned to quick-march back to Silesia at the beginning of August to prevent the Austrians and Russians from uniting. At Liegnitz on August 15 it was on the right flank in the second line but did not get into the battle. When the corps moved to Saxony via Guben, it went back to Silesia in Goltz's Corps and was later in winter quarters between Hirschberg and Frankenstein. In 1761 Colonel von Pomeiske became its chief on April 9; the regiment remained in the King's Army, which was at Strehlen and in camp at Bunzelwitz. On August 15 it took part in the Battle of Wahlstatt. On September 12 it advanced through Gostyn to Far Pomerania under Platen, broke through the siege of Kolberg on October 2 and broke out again with two squadrons on October 17 to fetch ammunition and food. On October 25 one squadron used its weapons at Treptow. Three squadrons broke out along the coast on November 14 and tried in vain to relieve the city on December 12. On July 1, 1762 it marched to Adelsbach, Friedland and Trautenau in Wied's Corps, then behind the front to Leutmannsdorf.

In Württemberg's Corps in 1806, it surrendered at Ratekau on November 7. In 1914 it was Cuirassier Regiment 4.

9. Dragoner-Regiment

DRAGOON REGIMENT 10

Commanders of the Regiment

1741	4/16	Lieutenant General Hans Friedrich von Platen
1743	11/14	Colonel Johann Adolf von Möllendorff, later Lieutenant General
1754	9/14	Major General Friedrich Ludwig, Count Finck von Finckenstein, later Lieutenant General
1785	9/23	Major General Wilhelm Leopold von Rosenbruch
1790	9/25	Major General Carl Wolfgang von Franckenberg
1795	1/31	Major General Carl Gottlieb von Busch, later of Dragoon Regiment 8
1801	4/26	Colonel Christian Heinrich von Manstein, later Major General
1806	8/19	Major General Ulrich Leberecht von Heyking

The twin regiment of Dragoon Regiment 9 not only grew from the same roots, but also had a very similar development and life history as the second-youngest dragoon regiment. It also originated in the squadron of "light" dragoons with light, fast horses and lighter riders that was set up in Dragoon Regiment 1 on September 1, 1722 for special tasks, reconnaissance, securing and mobile use rather than for closed cavalry attacks in firm battle order. There was certainly a need for these tasks, though they were taken over by the Hussars later, and development was slow until they were seen more and more from 1740 on. Except for special troops, such as Kleist's, "light dragoons" were no longer needed then. Until that time they were increased as usual to two squadrons in 1729, five in 1734 and ten in 1740. Since its separation from its source, Regiment 1, on April 16, 1741 it was stationed in Schlawe, Lauenburg and Btow under its chief of sixteen years, General von Platen; then it served first in East Prussia, then in the southwestern part of the Mark Brandenburg close to the Saxon border north of the Elbe in the Observation Corps of Prince Leopold of Anhalt-Dessau; during the last half-year of the First Silesian War it was on the lower Oder, then back in the western Mark for securing and on reserve duty. Only when Platen's "light regiment" was separated on December 1, 1743, which brought about the normalizing of "heavy" dragoon regiments, did it come into its first full battle action at Kesselsdorf in the winter campaign of 1745. These uses also corresponded to its capability and usefulness, scarcely to any mistrust of its leadership or quality. For this regiment and Dragoon Regiment 9 did their duty, even if they had the good luck to have to fight and prove themselves mainly in secondary actions against Saxons and Austrians, Russians and French. Its first chief as an independent regiment, after Platen, was Major General Johann Adolf von Möllendorff, who served for eleven years beginning in 1743 after being chief of Cuirassier Regiment 9 from 1741 to 1743. He was followed by Major General Friedrich Ludwig, Count Finck von Finckenstein, who served for 31 years and was taken prisoner at Torgau when his horse was shot. His close relative Carl Wilhelm von Finckenstein (1714-1800) was the King's boyhood friend and schoolmate, state and cabinet minister, and intimate, influential advisor for more than fifty years.

The regiment's replacements from 1743 to 1764 came from the area around Riesenburg in East Prussia, from 1764 to 1797 from the districts of Neidenburg and Mohrungen with the cities of Ortelsburg, Neidenburg, Passenheim, Willenberg, Osterode, Mohrungen, Saalfeld, Hohenstein and Liebstadt, from 1798 to 1806 from the dictricts of Neidenburg, Willenberg, Ortelsburg and Soldau, plus Baranowo, Zaremba and Mycinicz and their cities in New East Prussia. The garrisons were the same as those of Dragoon Regiment 9 from 1741 to 1743, Riesenburg as of 1743, from 1746 to 1788 Osterode, Mohrungen, Hohenstein, Saalfeld and Neidenburg, the last replaced by Liebstadt as of 1780; from 1789 to 1791 Allenstein replaced Osterode, in 1796 and 1797 it was in Osterode, Mohrungen, Wormditt, Saalfeld and Liebstadt. From 1797 to 1800 it was in Praschnitz, Mlawa, Bialla, Szuczyn and Johannisburg, the last replaced by Mycinicz in 1798. Its garrison in 1801 was Osterode, from 1802 to 1806 also Hohenstein, Ortelsburg, Löbau and Strasburg in West Prussia. Here too it always was close to its twin regiment.

The two years between the two Silesian Wars also gave the regiment time for intensive fighting drills meant to heighten its fighting and penetrating power. The King always insisted on lengthening the duration of the attacks. The cavalry had to attack with sword in hand, not shooting, always attacking first and not being attacked: "attack with the greatest speed and force and try to outrun, attack the enemy in the flank and send him flying sooner". In 1744 the King removed troops from East Prussia, where no danger threatened, and drew all five dragoon regiments into the area between Brandenburg and Magdeburg because of Saxony. Whoever attacked Silesia could not get past Saxony, which was also keenly interested in Silesia, at least as a link with Poland. Still an enemy of Austria in the First Silesian War, it had now been allied with Austria since December 20, 1743. For Prussia, fighting its way up at that time, Saxony was a nearby and dangerous competing power that could threaten Berlin as well as Prussia's flank and rear in Silesia and Bohemia. On July 27, 1745 the King ordered Prince Leopold of Anhalt-Dessau to prepare his corps to march, and on August 31 it stopped at the Saxon border because the King wanted to put off an open break with Saxony. When the Austrians advanced to Queis at the end of October and the Saxon Grünne Corps covered their left flank and threatened Berlin, the time had come: at Kesselsdorf, on the right wing in the second line under Kessler, it went through the village and the ravines to the south and attacked the Saxon infantry. Counterattacks by the scattered Saxon cavalry were torn up and did not prevent defeat. The regiment had 29 dead and four wounded, the highest casualties in the cavalry. If the destruction of the Saxon army had succeeded, it presumably could not have resisted so long in 1756.

In 1756 it was in Lehwaldt's Corps, which had the job of protecting the province and was, though certainly not mobilized by the King as of June 21, made ready step by step, since this time East Prussia was not to be given up either. Only when Russia made a military alliance with Austria on January 22, 1757 did the situation become more critical. At the end of that June the first Russians appeared on the border. The army of Field Marshal Count Apraxin, gathered at Kowno, crossed the border at Wirballen and moved slowly southward. Lehwaldt dropped back to Wehlau and met the Russians in battle at Gross Jägersdorf on August 30. On the right wing under the Duke of Holstein it threw back the cavalry of the former Saxon General Baron Sybilsky. But the battle was lost in the bloody forest battle that led to a retreat. It lost five dead, 29 wounded and 19 horses. Soon after that the Russians moved away to pick up reinforce-

10. Dragoner-Regiment

ments. When Lehwaldt had followed as far as Tilsit, he received orders to march to Pomerania against the Swedes. The Swedes gave up their pledges and held only Stralsund and Rügen. At year's end it besieged Stralsund. After the "Allied Army" had failed in the west in 1757, Duke Ferdinand of Braunschweig, the King's brother-in-law, took command of it at the end of the year. He put not only artillery, columns, treatment, clothing and equipment at his disposal, but also the fifteen squadrons of Lehwaldt's Corps under Duke Georg Ludwig of Holstein-Gottorp with 1800 horses, including the regiment, since the army was very lacking in cavalry. On March 3, while marching to the Rhine, it put French cavalry to flight at Lauenau near Hamelin, conquered a standard, smashed several battalions and took ten officers and 186 men prisoner, itself losing only nine dead, eighteen wounded. Crossing the Rhine near Emmerich, it fell upon the Bellefond Regiment and took its kettledrums. Then it fought at Goch and Sonsbeck. At Krefeld on June 23 it attacked the French infantry from behind and drove them out of their fortifications after a stubborn fight. It captured the kettledrums of the "Royal Cravattes" Cavalry Regiment. Then at the end of September it covered the withdrawal across the Rhine at Bork. In 1759 it had to meet the advance of the French at Freiensteinau in North Hesse. On April 10 it was in Ulrichstein. In the unsuccessful attack on the French positions at Bergen-Enkheim on April 13 it lost eight men and five horses. On April 19 two squadrons were attacked and almost wiped out at Queckborn. In the victory at Minden on August 1, which risked a withdrawal and again brought relief, it lost twelve men and twelve horses. Near Detmold it attacked a supply train and captured 280 baggage wagons and the Saxon war chest. Then it fought successfully at Naumburg in Hesse and at Nordeck near Giessen. After the bitter year of 1759 the King ordered the regiment to Saxony because of the heavy losses of at least six cavalry regiments. There it made Duke Ferdinand's life miserable; it was not at all worthless. It fought at Dresden in July of 1760 and Liegnitz on August 15, on the western wing in the second line, then in Silesia. In Goltz's Corps in 1761, it fought in the Battle of Wahlstatt, near Strachwitz, on August 15, completely smashing two Austrian cuirassier regiments and the mounted grenadiers who hurried to the scene, taking two standards. Surrounded in the pursuit, it fought its way out with bare swords, losing 137 men, with forty missing. The King thanked it with three Pour-le-merite and a gift of 2900 Taler. On September 12 it advanced under Platen to Gostyn, where on September 15 it captured two cannon and took a Russian battalion prisoner. As of October 2 it fought in the Kolberg area. After the vain breakthrough at Spie on December 12, it still has 200 men. In 1762 it went through the turning action at Adelsbach and Friedland in Wied's Corps, leading to Kniggrtz on July 10, and then took part at Burkersdorf and Leutmannsdorf as well as in the siege of Schweidnitz.

After a particularly good job at Warsaw on August 26-28, 1794 and Poponsk on August 31, it was in Württemberg's Corps in 1806 and surrendered at Ratekau on November 7. In 1914 it was Cuirassier Regiment 4.

10. Dragoner-Regiment

DRAGOON REGIMENT 11

Commanders of the Regiment

1740 12/18 Colonel Christoph Ernst von Nassau, later Lieutenant General
1755 11/27 Colonel Christoph Ludwig von Stechow, later Major General
1758 3/6 Colonel Leopold Johann von Platen, later Lieutenant General
1770 9/15 Colonel Franz Gustav von Mitzlaff, later Lieutenant General
1778 3/11 Major General Friedrich Leopold von Bosse
1789 5/20 Major General Carl Wilhelm von Tschirschky
1793 11/5 Colonel Ludwig Ernst von Voss, later Lieutenant General
1806 8/19 Major General August Friedrich Erdmann von Krafft

As proclaimed on December 18, 1740, Colonel Christoph Ernst von Nassau, who came from Saxon service and became a Count in 1746, recruited men for a new regiment that was to have ten squadrons. The recruitment and formation took place in Lower Silesia, particularly in Crossen, Grünberg and Freystadt, and made such fast progress that Nassau, who soon became a Major General, had gathered 600 men on "light, compact horses" by the end of April 1741. Like Friedrich Wilhelm von Kyau, who also came over and later became a Lieutenant General, most of the officers came from Saxon service. By order of June 21, 1741 the strength of the regiment was limited to five squadrons. The formation took place right after the march into the Silesian border area that had not yet been pacified. Nassau (1686-1757) served as chief for almost fifteen years and gave the regiment its atmosphere, especially as he was a born Silesian. He had proved himself particularly as the leader of an independent company in the Second Silesian War—as on February 9, 1745 in a surprise attack at Ratibor, on June 4 on the left wing at Hohenfriedeberg, and in mopping-up duty in Upper Silesia and the Silesian mountains—he was accorded the Order of the Black Eagle in 1744 for his clever withdrawal from Kolin and the assimilation of the Prague occupying forces in the fall. His excellent book, "Contribution to the History of the Second Silesian War" (Leipzig, 1780) became famous. His successor, Christoph Ludwig von Stechow, was a more versatile performer. In 1741 he established the Brieg Garrison Regiment out of "Whitecoats", then he prepared uhlans, he led a detachment at Mollwitz, then became commander "en chef" of I.R. 12. From November 27, 1755 to March 6, 1758 he was chief of the regiment as a Major General. From 1741 to 1743 it had no canton but stayed at full strength through recruiting in Silesia; from 1743 to 1747 the Silesian infantry regiments supplied replacements. From 1747 to 1806 they came from the districts of Sagan and Grünberg with the cities of Sagan, Naumburg am Bober, Priebus, Wartenberg and Grünberg. Its garrisons in 1743 were Sprottau, Grünberg, Wartenberg, Beuthen/Oder and Freystadt, then various places in Upper Silesia. From 1746 to 1755 they were limited to Sagan, Grünberg, Beuthen and Sprottau; from 1764 to 1777 it remained thus, but as of 1779 Freystadt was added again, and from 1796 to 1806 Beuthen was omitted.

The regiment saw its first action in the taking of Neisse, which was surrendered on October 31, 1741 according to the Protocol of Klein-Schnellendorf of October 9, after the Austruan army withdrew to Moravia and there had been heavy bombardment. In 1742 it helped to occupy important places to protect the connections to southern Upper Silesia and northern Moravia. On March 5 a company under Captain von Froideville was hit hard by hussars and Wallachs at Napajedl on the March north of Ungarisch Hradisch, but defended itself bravely until it was relieved the next day by the regimental commander, Colonel Baron von Kyau. At the end of April 1742 it left for Bohemia under Major General von Derschau, but reached the King at Czaslau only after Chotusitz. On June 1 it was camped between Maleschau and Kuttenberg. When peace came on June 11, it went into garrison in Silesia, under the direction of General von Buddenbrock. In all there were now seventy squadrons with 11,000 men. The short peacetime was spent on more intensive training for battle. The King demanded two things to beat the enemy: "First of all, attack him with the greatest speed and force, and secondly, try to outflank him. Every cavalry officer must never let himself be swayed from the thought of trying to attack the enemy in the flank so as to beat him faster". In 1744 it marched in the King's First Corps to Prague, which surrendered on September 16, and advanced to Tabor and Budweis. In the first withdrawal to the other side of the Elbe, and in the second, back over the Silesian mountains, after the enemy's break-through as of November 19, it repeatedly got the army out of unfortunate positions under its chief, without losing many men or horses. On June 1, 1745, under Nassau, it maintained connections between the two parts of the army in Striegau and northwest of Schweidnitz behind the Nonnenbusch north of Zirlau. At Hohenfriedeberg it was on the right wing in Posadowsky's dragoon line, where it held off the Saxon attack, broke into the retreating grenadier corps and surrounded five and a half companies under Major de Froideville. After Colonel von Schönberg had refused the request to surrender it was cut down, Schönberg was killed, and the survivors were taken prisoner. Great bitterness prevailed against the Saxons at that time because in 1744 they had attacked the army in the back in Bohemia. At the end of July it went under Nassau to mop up Upper Silesia, at first in Neustadt, then before Kosel on August 26, which was taken on September 5 after numerous days of bombardment. In mid-October it secured the border between Jägerndorf and Troppau. In an advance toward Leobschütz it defeated a Hungarian corps and took 170 prisoners. The enemy was kept in suspense until peace was made on December 25.

In 1756 it belonged to the Second Silesian Corps of Field Marshal Count von Schwerin, which carried out the six-week apparent attack from the County of Glatz to the Elbe in mid-September, and went into winter quarters early in December. In mid-April of 1757 it marched to Bohemia, reaching Prague on May 2. In the cavalry battle it was on the extreme left flank in Zieten's cavalry reserve south of Sterbohol, leading the decisive flank attack from the south with Hussar Regiments 3 and 4 while Zieten encircled the enemy from the west. The enemy evacuated the field. Colonel Georg Friedrich von Winterfeld, the regimental commander, fell. Under the Duke of Bevern it took up the pursuit, which led in the direction of Kuttenberg. When Daun tried to cut him off, he withdrew smartly toward Kolin. In the battle it was on the left wing behind the infantry of Hülsen's Division which, as the avant-garde, was to attack the dominating heights and the oak forest via Krzeczhorz, but was unable to take the heights but held the village and the woods to the end. Despite the defeat, the King could rebuild the striking force of his army.

11. *Dragoner-Regiment*

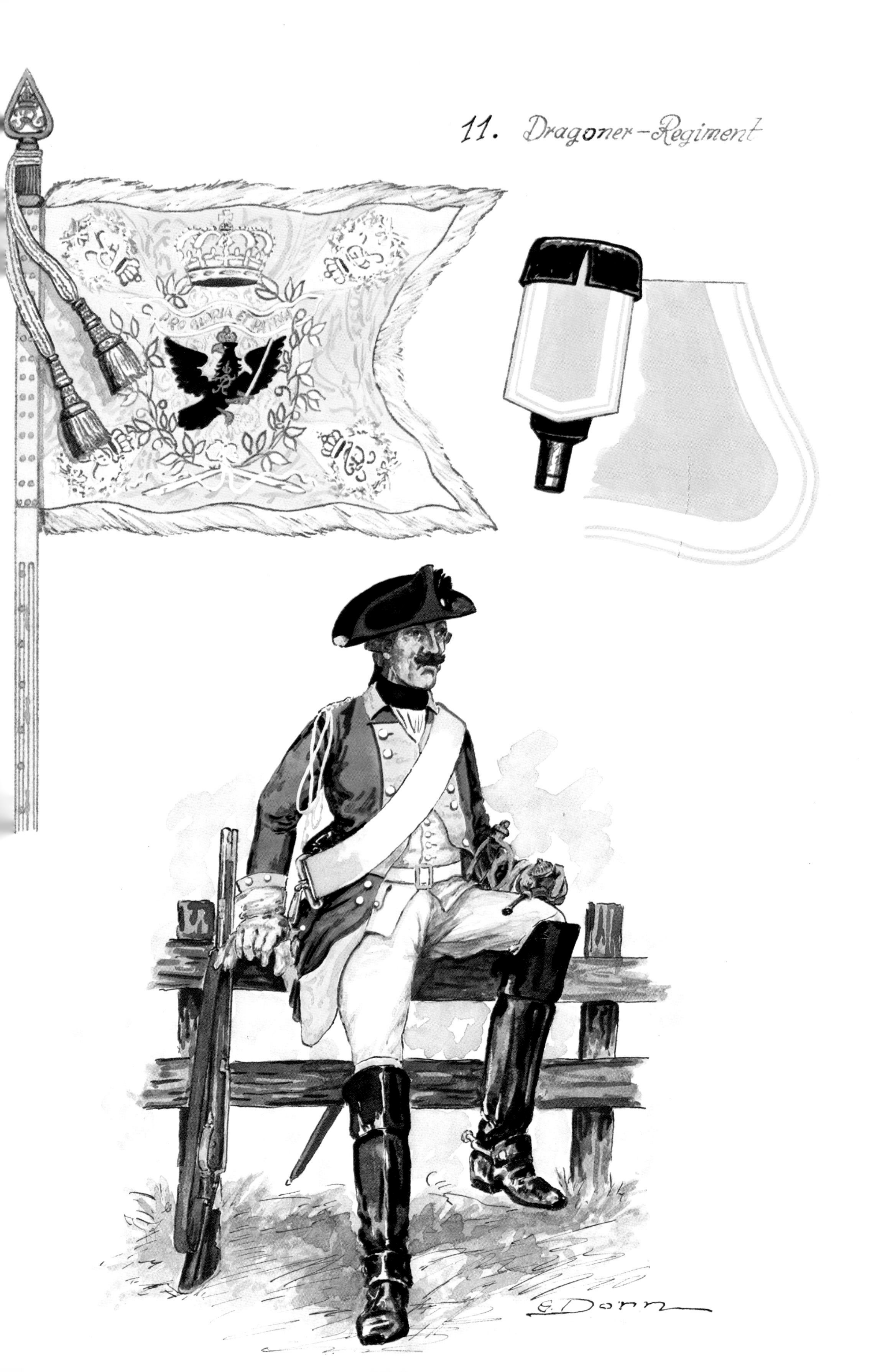

The regiment went to Jungbunzlau with the Prince of Prussia, then in August it went with Bevern from Zittau to Silesia. After the Battle of Moys on September 7 it marched through Bunzlau and Steinau into its position along the Lohe, which was broken through on November 22 by superior enemy forces. On December 2 it met the King's Army, which had hastened there, at Parchwitz. At Leuthen it was on the right flank in Zieten's cavalry corps in the second battle line, next to Dragoon Regiment 2 under Krockow, which first struck Nadasdy's cavalry, then the Austrian cavalry flank, and finally drove the infantry from the field, taking many prisoners. Among the baggage train it captured ninety flour wagons. With Zieten it pursued the enemy to Landeshut.

On March 6, 1758 Leopold Johann von Platen, whom the King had promoted from major to colonel on the battlefield at Kolin on June 18 as Commander of Dragoon Regiment 1 and valued very highly ever since, became the regiment's chief for twelve years. After the taking of Schweidnitz it advanced with the King's Army via Neisse to Troppau, where it was attacked and lost several hundred men. In the attack at Domstadtl on June 30 only a part of the regiment saw action. Then it went back via Königgrätz to Silesia, where it remained under Margrave Carl. Back with the King since September, it joined Retzow's Corps at Weissenberg, which neglected to occupy the Strohm-Berg, so that at Hochkirch it could only prevent a catastrophe and safeguard the retreat. At the end of October it had to go to Neisse to relieve it. In 1759 it left the Schmottseiffen camp with the Württemberg detachment and went from Bunzlau to Sagan to join Prince Heinrich and prevent the Russians and Austrians from joining forces. On the way to the Oder as of July 30, two squadrons drove back the Würzburg Regiment at Sommerfeld and took hundreds of prisoners. At Kunersdorf it fought in squadrons on the left flank behind the infantry, penetrated to the Judenberg and was last to leave the battlefield. In September it went to Torgau with Finck, proved itself well at Korbitz, was on hand at Strehla, and under Wunsch at Pretzsch on October 29, supported by amounted battery, smashed the whole enemy advance guard. For this day's work it was honored with the Grenadier March and Pour-le-merite medals for all nine staff and other high officers. On November 21 it was taken prisoner at Maxen after a brave fight. In Prince Heinrich's Corps in 1760 it had only two squadrons. During the Battle of Liegnitz it stayed near Breslau, having marched up from Landsberg. At Hermannsdorf it joined the King, who had arrived before Wittenberg via Guben on October 23. At Torgau, along with Dragoon Regiments 5 and 12 under Finckenstein in the second line, it successfully attacked the Süptitz Heights, took many prisoners and a standard. In 1761 its ranks were filled again and it played an active defensive role on the Mulde in the Saxon Corps. In Saxony in 1762 it experienced the breakthrough across the Mulde, camped at Pretzschendorf and served defensively on the Wild Weisseritz. At Frankfurt it joined the Jung-Stutterheim Brigade and moved against the northern part of the Spittelwald, put a dragoon regiment to flight and took a battalion prisoner before it reached Freiberg. In 1779 the King praised it as one of four regiments "distinguished" by good achievements. In 1806 it became part of Hohenlohe's Corps and was disbanded in the capitulation of Prenzlau on October 28. The remaining men joined the First West Prussian Dragoon Regiment.

11. Dragoner Regiment

DRAGOON REGIMENT 12

Commanders of the Regiment

1734		Duchess Marie Auguste of Württemberg, as of 1737 the Duke's Widow and Regent
1749	6/8	Colonel Friedrich Eugen, Duke of Württemberg, later Lieutenant General, finally Field Marshal
1769	7/9	Colonel Carl Erdmann von Reitzenstein, later Major General
1780	6/4	Major General Johann Nikolaus von Kalckreuth
1793	11/5	Major General Carl Gustav von Biberstein
1797	9/12	Colonel Carl Friedrich von Brsewitz, later Lieutenant General
1806	9/24	Colonel Friedrich von der Osten

As prearranged on September 28, 1741, King Friedrich II of Prussia took over from Württemberg—as he previously had done with an infantry regiment—the "Duchess Marie Auguste" Cuirassier Regiment founded in 1734 and reorganized in 1741 as the "House Dragoon Regiment" of the Duke's widow. It had taken part in the Imperial war against France in 1734-1735, fighting in Austrian service on the Rhine. Now it was only 300 men strong and was to be doubled by recruits from Württemberg. The thrifty, energetic Duchess held the chief's position as the widow of the late Duke Karl Alexander (1864-1737), since her sons were still minors. In 1749 she gave it to her third son Friedrich Eugen, after her eldest son Karl Eugen had become the reigning Duke. Since the regiment had been newly uniformed only shortly before, the King ordered on April 1, 1742 that it should keep its uniforms two more years in order to cover the cost of missing horses and equipment. On June 14, 1742, after being taken over by Adjutant-General von Kalnein under its commander, Colonel Konrad Leberecht, Marshal von Biberstein, it arrived in Berlin with two mounted and three non-mounted squadrons and was organized in Halle. Then it passed through Far Pomerania into the former quarters of Dragoon Regiment 2 in and around Treptow. It was now called "Alt-Württemberg". Its chief from 1749 to 1769, Duke Friedrich Eugen, had grown up—with his brothers—at his uncle's royal court since the age of nine and gone to his regiment at eighteen, which naturally soon led to difficulties. When the commander and officer corps no longer wanted to put up with their youthful chief's impetuosity, the King named Lieutenant Colonel Friedrich Wilhelm von Seydlitz its commander; in five month she disposed of the problems, then became the commander of Cuirassier Regiment 8. The Duke later developed into a proficient corps leader who gained the King's full recognition. On May 6, 1795 he became a Field Marshal, and on May 20 of that year he became the reigning Duke. His son Wilhelm Karl, born in Treptow/Rega in 1754, became Grand Duke in 1797, Elector in 1805 and King of Württemberg in 1805. From 1743 to 1794 replacements came from the districts of Bütow, Lauenburg and parts of Rummelsburg with the cities of Schlawe, Lauenburg and Bütow; from 1795 to 1806 from the South Prussian districts of Peisern, Brzesc, Radziejow, Wartha, Kalisch, Konin and Kowal. Its garrisons were as follows: from 1743 to 1772 Treptow/Rega, Wollin, Greifenberg, Massow and Naugard, from 1773 to 1774 Hohensalza (Inowraclaw) from 1775 to 1784 again in the Pomeranian areas, from 1785 to 1789 in Greifenberg, Reetz, Naugard, Wollin and Massow, from 1796 to 1799 in Kosten, Koschmin, Schmiegel and Karge, plus Krotoschin and Peisern at times, and from 1800 to 1806 in Kosten, Meseritz, Schmiegel, Peisern and Krotoschin.

In 1743 its optimal strength was 32 officers, 60 non-commissioned officers, one kettledrummer, four oboists, fifteen drummers, 660 dragoons, five flagsmiths, twelve lower staff members, 60 supernumeraries and 745 horses, not counting the officers' horses, in all 849 men. After a few months it was changed to Prussian drill and training, particularly the new regulation: "The cavalry achieved quickness and smoothness in its movements, and so all parts of the army worked with equal zeal on the full development of that discipline that made victors of the Romans", the King judged in 1746. Preparation for the campaign began as early as February of 1744; the regiment belonged to the King's First Corps, which marched through Saxony to Bohemia in the latter half of August and took Prague after a short siege. The advance to southern Bohemia exposed the lines to the rear and finally led to a retreat, which ended in the Silesian mountains late that year. In the process it lost quite a number of units when the Württembergers deserted. In 1745 its restoration was almost finished by April. At Hohenfriedeberg on June 4 it attacked at Thomaswaldau as the reserve of the second line on the left wing under Zieten, joining his regiment when Kyau's Cuirassier Brigade broke away and, in the rough terrain, had to utilize the place to cross at Teichau when Zieten found it. So it led Nassau's victorious attack which opened the way for the infantry. After that it held a position on the Elbe at Königgrätz in northern Bohemia for three months. Then it was in camp at Staudenz when the King, on the morning of September 30, unexpectedly found himself facing a decisive battle. During the battle at Soor it and six grenadier battalions under Major General von Schlichting covered the rear of the army against Nadasdy's Corps, whose attack would have been catastrophic. The King knew that only the army had saved him here. Except for operations in Saxony and mopping up the border areas, the campaign was over.

In the King's Army it crossed the Saxon border on August 29, 1756. As of September 10 it enclosed the Saxon army at Pirna, under Prince Moritz of Anhalt-Dessau, while the King fought the Austrian relief troops in the Battle of Lobositz. On October 16 the Saxons capitulated; the King had many reasons for wanting to include them in his army. So the Rutowsky Chevauxlegers Regiment with its four squadrons of light dragoons, each with six officers, eight non-commissioned officers, two drummers, one flagsmith and a hundred privates, was added to the regiment as its second battalion, even keeping its Saxon uniforms. When the new advance began in the spring of 1757, many Saxons deserted in March; the rest were divided in April. The integration had failed. In the breakthrough at Reichenberg on April 21, 1757 it joined Dragoon Regiment 1 in capturing three standards and several cannon after driving the enemy cavalry back immediately. Then it moved to Prague. On May 6 it was in the second line of the left wing south of Sterbohol, between Dragoon Regiments 1 and 2 and attacked when the cuirassiers dropped back past the village. The cavalry battle moved back and forth several times; threatened from the rear by Zieten's flank attack, the enemy cavalry turned and fled. On May 11 Lieutenant Colonel von Münchow and Major von der Trautenburg received the Pour-le-merite. The losses were

12. Dragoner-Regiment

high; the commander, Colonel Friedrich Wilhelm, Prince of Holstein-Beck, had fallen. Obviously it was not in action at Kolin six weeks later, but was used otherwise. After the siege of Prague ended on June 19 it moved via Brandeis and Jungbunzlau to Bautzen, and from there to Silesia under the Duke of Bevern. At Moys on September 7 it came through Nadasdy's attack; then it came from the east, via Bunzlau, Liegnitz and Steinau, to take a fortified position on the Lohe west of Breslau, because the direct route was blocked. On November 22 the Austrians ended their eight-week defense in an attack with threefold superior forces, after four hours of artillery fire, to meet the King who had hurried there from Rossbach. In spite of all their bravery, the battle was lost. Six days later the King was at Parchwitz. At Leuthen the regiment marched in the avant-garde under its chief, now a Lieutenant General, in a feint between Borne and Gross Heidau, taking 600 prisoners and two standards from the enemies and preventing them from seeing through the march to the right. In the battle it was behind the infantry's center along with four hussar regiments. "It won a lot of fame through its bravery".

At Landeshut in 1758 it protected the siege of Schweidnitz, then it surrounded Olmütz with the King. When the siege was given up after four weeks and the King turned back to Silesia via Königgrätz and Landeshut to defend the Oder, it stayed there under Margrave Carl from August 10 on. As of September 10 it was back with the King, and in mid-October it was in Retzow's Corps at Weissenberg; then its chief hurried with it to help at Hochkirch and guarded the withdrawal. On November 7 it helped to relieve Neisse, and on November 20 it was among the troops that moved into Dresden. The King's instructions of March 16, 1758 required: "When the infantry already has made a hole", the cavalry must "go in with whole squadrons in columns, one after another, and profit from the enemy's confusion". From Landeshut it went to camp at Schmottseiffen, then left with Prince Heinrich as of July 29; after Kunersdorf he marched first to Sagan, then to Bautzen, and then to Saxony at the end of September. On September 25 it fought successfully at Hoyerswerda, on October 29 at Pretzsch under Rebentisch. Then under Finck it left Strehla on November 8 and advanced to Rosswein via Eilenburg and Döbeln. After two days of battle at Maxen, four squadrons had to surrender after vain attacks and high losses, including its commander. The remaining squadron joined two squadrons of Dragoon Regiment 11 in 1760—its canton was occupied by the Russians—to serve under Prince Heinrich at Landsberg, Glogau and Berteslau. Back with the king after Liegnitz, it moved to Torgau, where it attacked successfully under Finckenstein in Holstein's cavalry. At full strength again in 1761, it joined its chief's Pomeranian Corps in defending Kolberg. On November 14 it broke out at Treptow, turned around and took part in the luckless storming at Spie on December 12, then fought its way back to Stettin. In 1762 it fought at Borkersdorf and Leutmannsdorf, and on August 16 at Reichenbach it had great success on the right flank of Bevern's Corps under Lentulus, then took part in the siege of Schweidnitz until October 10. In Le Coq's Corps, then Blucher's in 1806 it was disbanded in the capitulation of Ratekau on November 7. The depot forces in Kosel were later divided, some joining Hussar Regiment 6.

12. Dragoner-Regiment

HUSSAR REGIMENT 1

Commanders of the Regiment

1740	7/7	Colonel Johann von Bronikowski, later Major General
1747	9/10	Colonel Henning Otto von Dewitz, later Major General
1750	10/2	Colonel Michael von Szekely, later Major General
1759	5/16	Colonel Friedrich Wilhelm Gottfried Arend von Kleist, later Major General
1767-	1770	vacant
1770	9/15	Colonel Georg Oswald, Baron von Czettritz, later General of the Cavalry
1797	1/126	Colonel Anton Wilhelm von L'Estocq, later General of the Cavalry
1803	5/15	Major General Ernst Philipp (von) Gettkandt

King Friedrich Wilhelm I sent the commander of Dragoon Regiment 6, Lieutenant General von Wuthenau, the means to recruit and outfit two companies of Hussars in Insterburg on November 1, 1721. In the War of the Spanish Succession the Old Dessauer, serving with Prince Eugene's army, had learned to appreciate the Hungarian hussars and their capabilities, and in 1715 he had asked the Prince for advice in setting up "light riders".

Officers and men, 170 of them, came mainly from Hungary. Their garrison was Memel, their commander Major Konrad Schmidt of Dragoon Regiment 6. As later in the artillery, bourgeois officers always had a chance for promotion. In September of 1724 Captain von Gabor's company deserted except for thirty men. He was handed over by the Russians and executed; the company was filled with smaller men and dragoons and turned over to Captain Johann von Bronikiwski, who had come from Polish service. In the division of Dragoon Regiment 6 on July 15, 1727, the "Prussian Hussars" went to Dragoon Regiment 7 of Dockum, then back to their old regiment in 1732. In February of 1730 the King had them expanded to three squadrons of 100 privates each, one in Tilsit, two in Ragnit. A major since 1728, Bronikowsky became the commander. As of March 1, 1734 the squadrons grew to 134 privates each, adding up to 464 men. From 1735 to 1737 they again were part of Dragoon Regiment 7. From August to October of 1739 they were doubled to six squadrons and stationed on the border between Ragnit and Lyck to guard against Polish bands and prevent deserters from escaping. According to the regulations of Prince Eugen of Anhalt-Dessau, Chief of the Corps, the hussars were not intended for closed attacks with drawn weapons between 1736 and 1739, but to successions of swarm attacks with pistol and sword. By the end of May 1740 the corps included 24 officers, six ensigns, an adjutant, a quartermaster, 4 non-commissioned officers, twelve trumpeters, three medics, three flagsmiths and 720 hussars, a total of 825 men with 795 horses. After Captain the Count zu Dohna's squadron was transferred to establish Hussar Regiment 3, the remaining five were put under Colonel von Bronikowski as an independent regiment. It was known as the "Green Regiment". The most prominent figure among its chiefs was Friedrich Wilhelm Gottfried Arend von Kleist (1725-1767). After studying at the Knights' Academy in Brandenburg and the University of Halle, he joined Cuirassier Regiment 10 at the age of twenty and became a Lieutenant in 1753. In 1756 he was promoted from Lieutenant to Major—a sign of zealous advanced training as well as military ability—and transferred to the regiment, becoming its chief in 1759 as a Colonel. He used it as the nucleus of his famous Free Corps in the Seven Years' War, winning praise again and again from the King and Prince Heinrich for his illustrious achievments, and becoming one of the most highly trained Prussian officers. Its replacements came from Dragoon Regiments 2 and 11. Its garrisons were Goldap, Ragnit, Stallupnen, Oletzko, Pillkallen, Schirwindt, Lyck and Tilsit from 1735 to 1739. In 1743 it was in Guhrau in Silesia. From 1746 to 1783 it was in Steinau and Sulau, plus Winzig, Wohlau, Prausnitz, Schlawa and Herrnstadt from 1746 to 1755, Trachenberg from 1747 to 1783, also Ohlau, Winzig, Koeben. Schlawa and Stroppen from 1764 to 1783. From 1784 to 1791 it was in Sulau, Herrnstadt, Koeben, Tschirne, Trachenberg, Guhrau, Schlawa, Winzig, Steinau and Beuthen/Oder, in 1792 and 1793 in Wohlau instead of Beuthen, from 1796 to 1799 in Militsch instead of Herrnstadt, Tschirne and Schlawa, in 1800 Herrnstadt and Prausnitz as well. From 1801 to 1806 its garrisons were Wohlau, Guhrau, Herrnstadt, Militsch and Koeben, from 1804 on also Trachenberg, Sulau, Prausnitz, Steinau and Winzig.

In Jaunary of 1741 it sent three squadrons to Silesia. In return, three new squadrons were formed of veterans and recruits in East Prussia, so that it had eight squadrons, five of them in East Prussia. In 1741 the squadrons in Silesia were enlarged to 150 privates each, and that of Major von Mackerodt was ordered to join the Observation Corps in camp at Göttin early in April. Early in June the Silesian squadrons and those of the Bodyguard Corps formed Hussar Regiment 2 under Zieten. At year's end the regiment took part in the capture of Neisse, then went to northern Bohemia under Hereditary Prince Leopold. The King always sent hussars taken over from Hungary to the Göttin camp, enlarging Mackerodt's squadron to two, then to five. On August 9 they became independent as Hussar Regiment 5. On September 24 the King ordered all hussar regiments raised to ten squadrons; recruiting began in January of 1742 in the Imperial cities on the Rhine and in southern Germany. Early in February of 1742 two squadrons went to Olmütz with the King. In instructions of March 21 the King asserted: "The officers shall train the men just as in the dragoon regiments, and at all times remind their men that they must attack closed most of the time and with saber in hand." Against hussars, one column at most was to swarm out. Thushussars were also usable in battle. At Chotusitz, having just arrived from Kuttenberg, it joined Buddenbrock's attack on the left wing and smashed the Thüngen Regiment in the second battle line. It had come through its baptism of fire. In 1744 it marched to southern Bohemia in the King's First Corps. On February 6, 1745 it successfully attacked pandurs in Radun, southeast of Troppau. Margrave Carl's outbreak from Jägerndorf to Neustadt on May 22 included eight squadrons; at the same time, the others were part of Winterfeldt's attack south of Landeshut. Five squadrons were behind the second line in the center at Hohenfriedeberg but did not get into battle. As of June it mopped up Upper Silesia under Nassau.

In 1756 it moved via Aussig and Tetschen, where two squadrons stayed, to Lobositz. Here it served on the right flank at the Homolka-Berg and skirmished to Sullowitz before joining in the second attack. In January and February of 1757 it was expanded to 135 men. In 1757 two squadrons destroyed the magazine at Pilsen and moved into Franconia, while the others

1. Husaren-Regiment
G. Dorn

enclosed Prague on the south. At Kolin on June 18, five squadrons under Zieten covered the left wing east of Krzeczhorz against Nadasdy's cavalry after marching on at the fore. Zieten faced superior numbers of cavalry but cleared the battlefield only at 9:00 P.M. At the end of June it was at Leitmeritz; on August 31 it left Dresden with the King and went to the Saale. On September 7 it took besieged Pegau alone; on September 17, under Seydlitz, it attacked the Imperial army at Gotha against fivefold superior forces, received three Pour-le-merite and, as the avant-garde of Prince Moritz's Corps on October 17, relieved Berlin against Hadik's raider corps. At Rossbach it secured the march of Seydlitz's cavalry behind the Janus Hill, covered his left flank in the attacks and took four cannon. The pursuit extended to beyond Erfurt. From Novbember 13 to 28 it marched from Leipzig to Parchwitz. At Leuthen its Second Battalion was in the avant-garde, then behind the center of the infantry. In 1758 the regiment was in the Saxon Corps, strengthened to 41 officers, 90 non-commissioned officers, ten trumpeters and 1300 men. It raided in Franconia and the Upper Palatinate and fought defensively on the Müglitz, where it received four Pour-le-merite.

In the summer of 1759 Kleist, its chief, created a squadron of "Free Hussars" of Hungarian turncoats who gave good service in the Franconian campaign. At Himmelkron, near Kulmbach, on May 11 it smashed the Riedesel Detachment. Then it hurried to the King via Bautzen and Sagan. On the march to the Oder, Kleist shattered enemy hussars at Sommerfeld, and at Markersdorf he seized Hadik's baggage train and took its defenders prisoner. At Kunersdorf its eleven squadrons fought under Württemberg on the left flank, which attacked in desperation in the narrows of the Deep Path. The hussars were the last to leave the battlefield. Then it secured northern Silesia against the Russians. In 1760 Kleist expanded the Free Hussars to two squadrons and provided four squadrons of "light dragoons" for the King's Army, which failed to take Dresden. When the army went to Silesia in mid-June, it stayed behind under Hülsen. On August 20 at Strehla, Kleist and his sixteen squadrons attacked from behind and wiped out the Zweibrücken Chevauxlegers Regiment with the Baranyay Hussar Regiment, took three standards and many prisoners. Hülsen reported: "The cavalry has worked wonders!" In spite of that, Saxony was lost. After the King arrived on October 23, Kleist drove back Brentano's cavalry at Torgau while on the march to Torgau and took fifteen officers and 312 men prisoner; in the march onto the battlefield he led Zieten's Corps with the regiment and free corps, and attacked in Zieten's cavalry. In 1761 all advances into Thuringia helped to gain recruits, horses and money. Kleist enlarged the Free Hussars to five squadrons, the "Light Dragoons" to eight including a mounted battery, set up a battalion of "Green Coats" and a corps of rangers with three companies of 100 privates each, a brigade that stood out in scouting and securing. In 1762 Kleist enlarged not only his regiment but also his dragoons and Free Hussars to ten squadrons apiece, the Croats to two battalions. In the summer they repeatedly attacked in Bohemia, until in September they were pushed back to Freiberg. There Kleist used the mass of the first column and eight squadrons of Hussar Regiment 1 to take Klein Schirma and St. Michael and advance on the Three Crosses. Three squadrons were in Forcade's fourth column. Three Pour-le-merite attested to their success. The regiment had 300 Prussians, 382 Saxons and 1029 'foreigners'!

In Hohenlohe's Corps at Jena in 1806, it surrendered at Anklam on November 1 with 697 men; others reached Ratekau and East Prussia.

1. Husaren-Regiment

HUSSAR REGIMENT 2

Commanders of the Regiment

1741 7/24 Colonel Hans Joachim von Zieten, later General of the Cavalry
1786 3/1 Colonel Carl August, Baron von Eben und Brunnen, later Lieutenant General
1794 12/29 Major General Friedrich Eberhard Siegmund Günthervon Goeckingk, later General of the Cavalry
1805 10/19 Colonel Wilhelm Heinrich von Rudorff, later Major General

On September 30, 1730 King Friedrich Wilhelm I ordered the formation of a 71-man hussar company of capable transfers from the cavalry and infantry. The Margrave of Ansbach's Hussar Escort had pleased him so well when he visited his daughter that he wanted such a troop "for his own service". It came on October 8, under the command of Lieutenant Colonel Egidius Arend von Beneckendorff and served—at full strength as of November 1—to protect the King while traveling, provide orderlies and carry messages, and "for pursuit in desertions". By March 1, 1731 it was doubled to form a squadron of 163 soldiers. The second company was put under Lieutenant Hans Joachim von Zieten, who had just joined it and soonbecame a Captain. The First Company was in Berlin, the Second in Beelitz. On October 1, 1732 the Third Company joined it, along with 70 privates, and as of October 1733 it had three squadrons of 134 privates, 464 soldiers in all. In 1735 a squadron of Zieten's combined Prussian and Berlin hussars took part in the Rhine campaign, in order to study the Hungarian hussars. Their purpose was achieved in observing the most difficult conditions of hussar warfare, and they came under fire for the first time. After Benckendorff was cashiered, Lieutenant Colonel Alexander Ludwig von Wurmb followed him as commander. Since June 1 of that year it has been called the "Royal Bodyguard Hussars", which had fifteen officers, a quartermaster, three cornets, thirty non-commissioned officers, six trumpeters, three medics, three flagsmiths and 402 hussars as of 1740, a total of 464 soldiers with 444 horses. As a particular bit of decoration, it carried twelve tigerskins for the first three officers of each squadron, a gift from Sophie Dorothea, the consort of Friedrich Wilhelm I, which were to be worn over the left shoulder. King Friedrich II supplied the other nineteen, including those for the regimental adjutants.

In November of 1740 three squadrons of Hussar Regiment I joined in the course of mobilization for the march into Silesia. Zieten (1699-1786) was its chief and outstanding figure for 45 years; he joined I.R. 24 at Spandau in 1714 and became an ensign in 1720. Discharged in 1724, he became a dragoon in 1726 and was given his third position with the hussars. He developed them into a mobile fighting force and was their most popular leader. He proved himself as a corps leader and in battle at Hohenfriedberg, Leuthen, Hochkirch, Liegnitz and Torgau. His chief instructor was Lieutenant Colonel von Baranyay, his pupils chiefly Belling, Günther, Werner and Göckingk. He was cordial and God-fearing, industrious, active and sober, not a blind follower. The King said in 1760: "What a remarkable man this good Zieten is!" The regiment's replacements came from Infantry Regiments 19 and 25, from its canton only as of 1802. From 1736 to 1743 it was garrisoned in Berlin; the First Ballation remained there until 1787, the Second was in Parchim, Plau, Lübs and Eldena from 1743 to 1787, minus Eldena as of 1780. From 1788 to 1806 it was in Berlin, Fürstenwalde and Müllrose, as well as Beeskow as of 1803.

On December 16, 1740 it entered Silesia in the First Corps. In March its squadrons were enlarged to 150 men each. At Mollwitz on April 10 it was on the left wing in the second line after three squadrons had scouted at Pampitz-Neudorf and one had been assigned to guard the baggage train. When the Austrian cavalry under Berlichingen broke in, it helped to lead the counterattack under Zieten. On May 17 at Rothschloss Zieten attacked a dertachment under Baranyay with four squadrons, supported by two squadrons of Dragoon Regiment 5; the enemy barely escaped being taken prisoner. Zieten received the Pour-le-merite, and at the beginning of June the Berlin and Prussian squadrons were merged to make an independent regiment, of which he, now a Colonel, was the commander. "Under his leadership the hussars soon learned to handle their Hungarian enemies", Jany judged. By order of September 24, all hussar reguiments were to be enlarged to ten squadrons during the winter. In February of 1742 it went to Olmütz with the King; in mid-February it advanced to Stockerau and Korneuburg, near Vienna. Small engagements showed the value of the hussars. Every regiment received 100 trained carabiners. Early in May it went to Bohemia under Derschau but did not join the King until after Chotusitz. In 1743 the King sent 26 cavalry officers to the Hungarian army to recruit well-known cavalry leaders. On many a day up to ten hussar officers reported to the emissary in Vienna. On December 1 of that year there appeared a "regulation for the hussar regiments" which prepared for war experience. In 1744 it advanced to southern Bohemia with the King's First Corps. In the rear-guard action at Moldautheim on October 9 under Zieten it drove the enemy back. Until year's end, hussars and grenadiers secured the Silesian mountain line. On May 19, 1745 Zieten and four squadrons came from Patschkau to relieve Neustadt and within 22 hours broke through to Margrave Carl at Jägerndorf; on May 22 he began the withdrawal. On June 1 it secured the connections between the two parts of the army at Zirlau, under Nassau. At Hohenfriedeberg, when Kyau could not advance toward Thomaswaldau on the left wing, Zieten and the regiment found a place to cross at Teichau. On November 23 at Kath. Hennersdorf Zieten smashed three Saxon cuirassier regiments and the Saxe-Gotha Infantry Regiment, taking 1080 prisoners, so that Lusatia became free. During Kesselsdorf the regiment secured the area east of the Elbe.

On August 29, 1756 it marched to Pirna in the King's Army; two squadrons were at Dux. The first fatality was a Zieten hussar on September 21. At Prague in 1757 Zieten attacked from the south at Sterbohol, coming via Bechowitz during the cavalry battle, while the regiment fell upon the enemy's rear farther to the west and drove them away. The infantry fought out the victory. At Kolin it was west of Krzeczhorz behind Treskow but saw little action. Zieten was wounded in the head. On July 7 it was strengthened by fifteen cornets, thirty non-commissioned officers and 300 privates to 1500 horses, having practically two squadrons. When the King evacuated Bohemia and moved westward, it went to Silesia under Bevern. On September 7 it fought bravely at Moys, as also in the large-scale Austrian attack on the Lohe position before Breslau on November 22, where it fought on the left wing. A week after the King's return, it took Neumark in a surprise attack on December 4. It marched to Leuthen in the avant-garde, then attacked the Württembergers and Bavarians in the third line on Zieten's wing and took over 2000 prisoners. When Cornet von Quernheim and thirty men captured 1800 men and four cannon in the pursuit, the King

2. Husaren-Regiment

gave him the Pour-le-merite, a hundred ducats and a promotion to captain. But Zieten asked that the Cornet should "wait his turn for advancement and be satisfied with the medal and the gift". Quernheim became a lieutenant only at the end of 1758. It was the strongest regiment that year; it went with the King via Schweidnitz to besiege Olmtz, which the King broke off, and then marched to the Oder. At Zorndorf it attacked the Russian right flank "with extraordinary courage" and led the counter-attack against Demiku's storming. After the battle it had to leave for Beeskow at once to reconnoiter the Austrians in Lusatia, whither the King followed in two weeks. At Hochkirch on October 14 it was first in the saddle at the Schlosserschenke and threw itself into the battle repeatedly. "You could not get rid of them", Prince de Ligne wrote. Colonel von Seel, its commander, lost his life. Then it went to Silesia to relieve Neisse. At year's end it was in Saxony. In mid-June of 1759 the First Battalion under Major von Reitzenstein went from Glogau to Dohna's Corps on the Warthe, while three squadrons under Württemberg went to Bunzlau to join Prince Heinrich. At Kay on July 23 Reitzenstein's attacks could relieve the infantry but not change the outcome. At Kunersdorf six squadrons attacked on the left wing in the first line without being able to gain ground. When the King was in danger of falling into the Cossacks' hands, Captain von Prittwitz and his squadron rescued him. Then it covered Glogau and hurried to Saxony. On October 29 two squadrons under Wunsch led the successful fight at Pretzsch. In 1760 it was under the King before Dresden, and at the Katzbach as early as August 9. At Liegnitz Major von Hundt, with two battalions, was first to report the enemy's approach. It threw itself at the enemy on the left wing to win time to turn the front; when the enemy dropped back, it penetrated far in. Zieten became General of the Cavalry. After Hohgiersdorf and the relief of Berlin it went to the Elbe. At the head of the avant-garde at Torgau it smashed the St. Ignon Chevauxlegers Regiment and captured twenty officers and 400 men. Its commander, Major Zedmar, fell. In Holstein's attack in the afternoon it captured six flags. Hussar soldiers gathered up scattered cavalrymen and brought them in. At Langelsalza on February 15, 1761 Prittwitz's First Battalion took the Saxon Guard and a grenadier battalion prisoner, along with three cannon and a flag. On April 2 it shattered Imperial troops at Schwarza and took the commander and 400 men prisoner, along with four cannon and two flags. At Saalfeld both battalions pursued a stronger corps through the Saale and brought back 900 prisoners with six cannon and two flags. Soon thereafter Major von Hundt, Commander of the Second Battalion, fell. One Pour-le-merite apiece expressed thanks. In 1762 it went into battle at Burkersdorf in Silesia, then to Saxony, where on November 7 at Spechtshausen, near Freital, the Second Battalion under Prittwitz fought in Wied's Corps, attacking the enemy avant-garde, capturing the battlement and taking over 600 prisoners and four cannon. Two captains earned the Pour-le-merite. In 1763 it consisted of 1224 Prussians, fifty Saxons and 255 'foreigners'.

In Weimar's Corps, then Blücher's in 1806, it capitulated at Ratekau on November 7. The depot men and the rest reached East Prussia, where at first two, then four squadrons were established. In 1861 Hussar Regiment 3 inherited the tradition.

2. Husaren-Regiment

HUSSAR REGIMENT 3

Commanders of the Regiment

1740	7/2	Colonel Friedrich Asmus von Bandemer
1741	9/19	Colonel Hyazinth Malachow von Malachowski, died of wounds 4/17/1745
1745	4/20	Colonel Hartwig Carl von Wartenberg, later Major General, died 5/2/1757
1757	7/31	Colonel Carl Emanuel von Warnery, later Polish General
1758	3/27	Colonel Christian Möhring, later Major General, raised to the nobility in 1773
1773	5/5	Colonel Stephan von Somoggy, Regimental Commander since 1763
1777	12/7	Colonel Hans Christoph von Rosenbusch, later Major General
1785	9/18	Colonel Carl Franz von Keoszegy, later Major General
1788	5/23	Colonel Georg Ludwig Egidius von Khler, later General of the Cavalry
1796	6/24	Major General Friedrich Ludwig von der Trenck
1797	9/17	Colonel Dietrich Wilhelm Schultz, later Major General, raised to the nobility in 1798
1803	11/19	Colonel August Wilhelm von Pletz, later Major General

After the burial of his father, Friedrich II wrote to Voltaire that he would establish five squadrons of hussars as the first closed regiment, aside from nine squadrons he took over. It was formed around Captain the Count zu Dohna's squadron of the "Prussian Hussar Corps" with 120 privates each, 695 soldiers in all. The units took in dragoons and new recruits to fill their ranks by December 1. Lieutenant Colonel von Bronikowski, Commander of the "Prussian Hussar Corps", took charge of the formation for Colonel von Bandemer, who was in Polish service in 1713, Russian in 1719, and a Prussian Colonel in 1738. Early in April of 1741 one squadron joined Major von Mackerodt at the Göttin camp and then went to Hussar Regiment 5 as of August 9. On the march to the army, the new regiment was lured into an ambush near Leubus Abbey on August 1 and driven into the Oder. It still had three weak squadrons with six officers, seventeen non-commissioned officers, three trumpeters, 207 privates, five medics, three flagsmiths and 279 horses. Bandemer was discharged. The Austrian hussars were superior in adventurous spirit, mobility and capability. The King tried to equalize the situation by instruction and training as well as by enlarging the Hussar ranks. Its next chief, von Malachowski, came from French service. This splendid officer accidentally received a fatal wound in battle on April 12, 1745 from one of his own hussars. Its best-known chief was von Wartenberg (1711-1757), a cadet in 1725, a Lieutenant in 1731. With comrades such as Winterfeldt and Christoph Hermann von Manstein, he went to Russia as an instruction officer under Field Marshal Count Münnich, until the King called him home in 1740. In 1741 he joined the rgeiment, a highly capable Lieutenant Colonel, and during ten years of peace he made it into an exemplary regiment, to which the King regularly sent officers for training. A Major General in 17512, he fell—irreplaceably—at Alt-Bunzlau on May 2, 1757.

For fifteen years its chief was von Möhring, a man of bourgeois origins, who had served with the Berlin Hussar Corps and become a Corporal in 1735, Cornet in 1740, Captain in 1750 and Colonel in 1758. Not only well-educated young burghers, but non-commissioned officers too could rise, as did Werner, Gröling, Schultz, Hohenstock, Gettkandt, Rudorff, Pletz and Göckingk. Replacements were supplied by Cuirassier Regiments 1 and 9, from 1798 on Regiments 1 and 8. In 1740 its garrisons were Goldap, Lyck, Johannisburg, Sensburg and Ortelsburg, in 1743 Bernstadt and Gleiwitz, from 1746 to 1752 Gross Strehlitz, Bernstadt, Kreuzburg, Konstadt, Guttentag, Reichenthal, Namslau and Lublinitz. From 1753 to 1755 it was in Bernstadt, Buchwalde, Sadewitz, Korschlitz, Neudorf and Zellin in the country, from 1764 to 1795 in Kreuzburg, Pitschen, Tost, Bernstadt, Konstadt, Reichthal, Landsberg, Rosenberg, Guttentag and Lublinitz, without Landsberg from 1779 on. From 1796 to 1806 the garrisons were Bernstadt, Konstadt, Oels, Trebnitz, Wartenberg, Reichthal, Festenberg, Juliusberg and Rosenberg, as of 1800 with Medzibor in place of Rosenberg. The regiment was divided by squadrons.

At the end of October 1741 it was at full strength again and went to northern Bohemia with Hereditary Prince Leopold. In 1742 it was strengthened to ten squadrons when it was transferred to Upper Silesia. A genuine hussar war took place on the right side of the Oder. When peace was made on June 11, 1742 it stayed in Silesia. In 1743 its optimal strength amounted to thirty-six officers, eighty non-commissioned officers, ten trumpeters, ten flagsmiths, ten medics, and a six-man regimental staff. In 1744 it was in Upper Silesia in the Second Corps under General von der Marwitz. The corps consisted mostly of Silesian troops. It did not make the planned march to Olmütz because of attacks from Solvakia. It guarded Glatz, Jägerndorf and Troppau; otherwise small-scale war prevailed. On December 11 Wartenberg and three squadrons attacked "a considerable corps of cavalry" in Pless, created "a massacre that had a horrifying effect", and took 200 prisoners. At year's end it went back to Kosel and Neisse. In Nassau's surprise attack against Trenck's Corps in Ratibor on February 9, 1745 the regiment stood out: "I cannot sufficiently praise the bravery of the hussars, especially those of Malachowski, as the Colonel himself led the way and pursued the enemy as far as the bridge, had the hussars dismount and charge on foot until the infantry could follow," Nassau reported. Malachowski and Wartenberg received the Pour-le-merite. With 700 men under Hautcharmoy and Winterfeldt it beat the Hungarians at Ujest and Gross Strehlitz on April 12, at Würbitz west of Konstadt on April 20, at Gross Wartenberg on May 4 and Kreuzburg on May 18. As of the end of June it went to Neustadt, besieged Kosel and defeated a dragoon regiment at Oderberg in mid-October, taking a standard and over a hundred prisoners. After evacuating Upper Silesia in mid-November, it attacked the Philibert Dragoon Regiment at Schwarzwaldau near Landeshut on December 6, captured three standards and brought in many prisoners.

As the regulations of 1743 had stated that bourgeois non-commissioned officers with "great merit" could be nominated for promotion to lieutenant after twelve years of service, the hussar regulations said: "When officers leave, the commander shall suggest to His Majesty as officers the capable non-commissioned officers who apply themselves most to service and merit it, without regard for their status, apolitically according to the length of their service". In 1756 it advanced to Aujezd in mid-September as the avant-garde of Schwerin's Silesian Corps and drove back the enemy with some losses. At the end of October this pretended attack ended. In 1757 it marched to Jungbunzlau under Schwerin. At Altbunzlau on May 2 it smashed 1500 Croats from Königsegg's rear-guard but lost Wartenberg, mourned by the King as "one of the best officers". On the left wing under Zieten at Prague, it attacked Hadik's hussars in the flank three times from the south, driving the entire wing away. At Kolin it followed Hülsen's avant-garde over Krzeczhorz on the left wing and protected its left flank at Eichenbusch. It held the important position until evening, then secured the retreat. When the army was divided at the end of

3. Husaren-Regiment
G. Dorn

August, it marched to Silesia via Liegnitz and Breslau, taking the position on the Lohe. In the surrender of Schweidnitz on November 12 seven squadrons and their regimental chief Warnery were taken prisoner. On the march to Leuthen its three remaining squadrons formed the rear-guard and protected the baggage train. In the battle they stood behind the center and then took up the pursuit. Filled to 41 officers, 90 non-commissioned officers, ten trumpeters and 1140 privates in 1758, it went via Schweidnitz to enclose Olmütz until the eginning of August, and on the march back it beat a corps of enemy cavalry at Landskron. Under Wedell in September it advanced from Saxony to protect Berlin from the Swedes and beat them back to Fehrbellin at Linum on September 22, taking the town on September 28. After securing at Zehdenick and Templin it reached Torgau on November 12, at the right time to deflect Hadik before the Imperial army arrived. In 1759, strengthened to 100 non-commissioned officers and 1260 privates, it went from Schmottseiffen via Sagan to Kunersdorf, where three squadrons in Finck's Corps on the right wing led the initial attack over the Trettin Heights to the narrows between the Kuh-Berg and Elsbusch. With the King in Saxony as of October, it fought at Pretzsch on October 29 under Wunsch, driving Arenberg's avant-garde to flight, and winning a Pour-le-merite.

In 1760 it was with the King before Dresden, the siege of which he lifted to hurry to Silesia early in August. At Liegnitz it was on the right wing in the secondline, which did not come into battle. It was at Hohgiersdorf on September 17 but stayed behind when the King hurried to Berlin on October 7. Under Goltz it advanced to Grottkau early in November and drove back the Wolfersdorff Detachment at Landeshut. In 1761 it was expanded again to 110 non-commissioned officers and 1360 privates, and went via Pilzen to the camp at Bunzelwitz, and from there to Strehlen. In 1762 it had 51 officers, 120 non-commissioned officers and 1450 privates. At the end of July it fought at Burkersdorf-Leutmannsdorf; in mid-August it was in action at Reichenbach under Werner and then at Neustadt and Ratibor. In 1763 it consisted of 1493 Prussians, 35 Saxons and only 62 'foreigners'.

In 1806 the First Battalion was in Rüchel's Corps, the Second in Weimar's, then both were in Blücher's, which surrendered at Ratekau on November 7, as one unit did at Hamelin. From the deopt troops there originated the Schill and Hellwig Squadrons, which freed 10,200 Prussian prisoners with fifty hussars on October 17. Later it became Hussar Regiment 6.

3. Husaren-Regiment

HUSSAR REGIMENT 4

Commanders of the Regiment

1741	7/23	Colonel Georg Christoph von Natzmer, later Major General
1751	2/3	Colonel Heinrich Sigismund von Vippach
1755	9/12	Colonel Georg Ludwig von Puttkamer, later Major General, died 8/12/1759 at Kunersdorf
1759	12/23	Colonel August Levin von Dingelstedt
1762	11/16	Colonel Balthasar Ernst von Bohlen, former Commander of Kleist's Free Hussars
1770	9/15	Colonel Carl von Podjurski, later Major General
1781	3/16	Colonel Eugen Friedrich Heinrich, Prince of Württemberg, later General of the Cavalry, in 1794 Governor of Glogau
1794	12/29	Colonel Carl Ludwig von Lediwary, later Major General and Chief of Hussar Regiment 10
1797	1/23	Lieutenant General Eugen, Prince of Württemberg (as 1781)

On December 19, 1740 Major General von Gessler, Commander of Mounted Regiment 4 in Rügenwalde, received the assignment of recruiting light squadrons in Poland, six for every 150 communities. Lieutenant Colonel Georg Christoph von Natzmer of the regiment took charge of the formation. Like the Polish hussars of the Seventeenth Century, the riders from Masowien—called "Wallachs" or "Tartars"—carried long lances, used by the uhlans later but unknown to the Hungarians. At the end of March the first three squadrons were in Landsberg/Warthe with 400 men, whom the rest joined in April. At the end of May all organizing and arming were concluded, so that they could march to join the army at Ziegenhals early in June of 1741. On June 23 the king assigned the regiment, which still wore Polish hussar uniforms and was called the "Hulanen Regiment", to Colonel von Natzmer as its Chief. The Secret War Chancellery called it "Natzmer's Hussar Regiment" since that February. In Olbendorf, not far from Grottkau, it first encountered its capable enemy on June 7 and suffered—without success—considerable losses. Practice and familiarity with the lances were lacking. The King had a better relationship with Prussia than with the Hungarians as a guarding power of Protestantism. In February of 1742 two squadrons went with his corps to Olmütz, four squadrons to Upper Silesia. On April 1 he wrote to Hereditary Prince Leopold: "The uhlans are not worth their bread". The regiment was just being restructured with ten squadrons, mainly unmounted and untrained, the mounted men used as despatch riders and orderlies and divided without supervision by the officers, and without their lances. Criticism of them referred to discipline, organization and capability of deployment at this point in time. When the King went to Bohemia at the end of April, it stayed in Upper Silesia under Prince Leopold of Anhalt-Dessau.

When peace was made on June 11 it could make up for its deficiencies in Silesia. When it was reclothed in Prussian uniforms, the King allowed on March 29, 1743 that two officers "shall not have the present Hulan uniforms made, to spare the cost, and wait until the regiment will begin to wear the new uniform", which led to its being called "The White Regiment". Natzmer (1694-1751), as diligent an instructor as Belling and Knobelsdorff, trained an outstanding, well-educated officer corps in a few years, including Captain Friedrich Wilhelm von Seydlitz until 1752. Puttkamer (1715-1759) entered Cuirassier Regiment 4 in 1732, Lieutenant in Hussar Regiment 3 in 1740, instructional and training officer under Wartenberg, Major in 1745, Lieutenant Colonel in 1753, Major General in 1757 with the notation in his commission: "Since on so many occasions, particularly in the present war, he has shown examples of his bravery", fell on August 12, 1759 at Kunersdorf. Replacements came from Cuirassier Regiments 1 and 8 as well as Dragoon Regiment 2, as of 1798 from Cuirassier Regiment 4. Its garrisons in 1743 were Militsch and Festenberg, then until 1778 also Trebnitz, Oels, Wartenberg, Stroppen, Juliusburg, Prausnitz, Brallen and Medziborm from 1779 on the same plus Auras, from 1783 to 1795 the same without Auras, Brallen and Stroppen. From 1796 to 1803 it was in Kempen, Ostrowo, Krzepicze, Wieruschau, Zorek, Grabowa, Boleslawice, Dzialoszin, Wielun and Pilicze in the Polish border area. From 1804 to 1806 the garrisons were in Namslau, Kempen, Ostrowo, Roesnberg, Wielun, Wieruschau, Boleslawice, Dzialoszin, Siewierz and Radomsk, half in Silesia.

Its optimal strength in 1743 was 36 officers, 80 non-commissioned officers, thirty trumpeters, flagsmiths, medics, 1020 hussars, not counting the six-man regimental lower staff. The regulations of December 1 of that year required riding proficiency above all else: "that a hussar be so adroit on a horse that when the horse is at a dead run he can pick something up from the ground in one hand and take off his cap with the other while in pursuit". The hussar must "wheel and turn his horse on a space as big as a Taler, as he wishes". Natzmer had hunters recruited for his regiment and sent ten to every squadron as the best people for reconnaissance. In 1744 it marched to southern Bohemia with the King's First Corps and fought bravely in the difficult retreat, in which it lost several deserters, some of whom returned after the general pardon. In April of 1745 it was at Schweidnitz to secure the mountain passes. Under Winterfeldt at Landeshut on May 22 it attacked Nadasdy's superior forces in the first battle line and drove them back across the Reichhennersdorf Mountains. Winterfeldt, who became a general here at the age of 38, praised Captain Friedrich Wilhelm von Seydlitz as "an officer who is not to be improved". In du Moulin's Corps it camped between Striegau and Stanowitz on June 1. At Hohenfriedeberg under Winterfeldt it was on the right wing before Pilgramshain, attacked Schliochting's Saxon light cavalry and took them prisoner. In the pursuit to Landeshut it captured a pair of kettledrums. After defensive action on the upper Elbe, Captain von Warnery's squadron smashed a raiding party at Greiffenberg on September 26, "whereby the Bosniaks, mixed in with the hussars, acquitted themselves uncommonly bravely", while three squadrons rode on patrol with the King and reported the coming of the enemy at Soor before he was surprised. After that he described having too few hussars with his army as one of his mistakes. At Gross-Strehlitz in late autumn it struck a detachment "with the greatest courage" and took 112 prisoners.

On August 29, 1756 it crossed the Saxon border with the King's Army. The first shot was fired on September 3, as Lieutenant Colonel von Warnery overwhelmed the mountain castle of Stolpen with a few hussars and took the occupying troops, including General von Liebenau, prisoner. He received the Pour-le-merite. From September 10 to October 16 it was utilized at Pirna. After strengthening to 135 men at the beginning of 1757 it numbered 41 officers, 90 non-commissioned officers, thirty trumpeters, flagsmiths, medics, and 1140 privates. On April 21 it took part in the breakthrough at Reichenberg. At Prague on May 6, as the indecisive cavalry battle raged on the left wing south of

4. Husaren-Regiment
G. Dorn

Sterbohol, it attacked the south flank of Hadik's hussars over Unter-Mechlup under its chief, while Zieten moved out farther, supported by Hussar Regiment 3 and Dragoon Regiment 11. The infantry battle still had to be won. It took the most prisoners in the pursuit. At Kolin it was in Zieten's Corps, protecting the left flank against Nadasdy's superior forces until evening along with four hussar and two dragoon regiments. After the orderly retreat from Bohemia until the end of August, it marched off under Bevern to protect Silesia. After the Battle of Moys on September 7 it went via Bunzlau, Liegnitz and Steinau to Breslau, from where it took up a position on the Lohe. When the Austrians broke through on November 22 it was on the left wing under Zieten, holding Nadasdy's Corps in check at Kleinburg, possession of which changed several times. On December 2 it joined the King at Parchwitz, then marched in the avant-garde to Leuthen, stayed on the left wing and rode in Driesen's final attack as the third battle line, coming out of the cover of the Radaxdorf Heights against the positions north of Leuthen, which led to panic. Its pursuit continued to Landeshut. As of July 1, 1758 it was strengthened by 170 soldiers and numbered 41 officers, 100 non-commissioned officers, ten trumpeters and 1300 privates, making it—after Hussar Regiment 2—the second-largest regiment. After the taking of Schweidnitz it experienced the fighting to surround Olmütz, especially the attack at Domstadtl on June 30. When the King moved against the Russians on August 10, it stayed in Silesia under Margrave Carl, but joined him again at Grossenhain on September 10. In Retzow's Corps at Hartha it caused Laudon's rear-guard the loss of 500 men, and it appeared on the left wing at Hochkirch with one battalion in time to avoid the worst. While marching back to Silesia, it repelled enemy cavalry at Ebersbach on October 26 and took 450 prisoners. On November 7 it helped to relieve Neisse.

In the Saxon Corps in 1759, it was called to bring two squadrons from Saxony and eight from Lusatia in mid-June and join Dohna's Corps on the Warthe, and as of June 24 it advanced into Posen as far as Obornik. On July 23, under Wedell, it vainly attacked the Russians on the Palzig Heights through the Zauche Valley. At Kunersdorf it secured the army in front during its laborious march onto the field and then attacked on von Seydlitz's left cavalry flank. In the cavalry battle Puttkamer met death at the head of the regiment. Then it secured Silesia against the Russians north of the Oder. With Prince Heinrich in 1760 it went to the Warthe, then via Glogau to Breslau and the Hünern camp, so that the Russians would not attack at Liegnitz. At Hermannsdorf it joined the King on August 29. When he marched away on October 7 and turned toward the Elbe because of the Berlin situation, he took two squadrons along, while the regiment went from Lubben to Silesia under von der Goltz to protect it from the Russians. With the Saxon Corps in 1761 it defended its position at Schlettau and on the Triebisch by advancing against the enemy, such as at Plauen, where the Second Battalion caught the rear-guard of the Imperial army, took many prisoners and captured three cannon. The Mulde remained the front. In 1762 it took part in the capable delaying battles in Saxony without standing out particularly.

In 1806 it fought in the main army at Jena and Zehdenick. The regiment reached East Prussia with 407 horses and formed Zieten's First Hussar Brigade, in 1808 the First Silesian Hussar Regiment, to which all its men came, later Hussar Regiment 4, whose tradition the King did not recognize.

4. Husaren-Regiment

HUSSAR REGIMENT 5

Commanders of the Regiment

1744	3/10	Colonel Joseph Theodor von Ruesch, later Major General, Baron since 1753
1762	5/9	Colonel Daniel Friedrich von Lossow, later Lieutenant General
1783	10/18	Colonel Carl August (von) Hohenstock, former Chief of Hussar Regiment 8, later Major General
1788	5/23	Colonel Friedrich Eberhard Siegmar Günther von Goeckingk, ennobled 1768, later of Hussar Regiment 2
1794	12/29	Major General Friedrich Wilhelm von Suter, later Lieutenant General
1804	12/20	Colonel Moritz von Prittwitz, later Lieutenant General

Since the beginning of 1741, one squadron each of Hussar Regiments 1 and 3 were camped at Göttin under Major von Mackerodt in the corps of Prince Leopold of Anhalt-Dessau. They were expanded from 120 to 150 privates with Hungarian hussars. During the summer many captured or deserted Austrian hussars joined Prussian service. The King sent them to the camp, where their numbers grew. On August 9 of that year he ordered the squadrons formed into a regiment of five squadrons, with 810 privates, 899 soldiers in all. It had no chief; Mackerodt, of Hussar Regiment 1, became a Lieutenant Colonel and its Commander in November. Atypically, it was given black uniforms, with a skull on its caps that came from Austrian pandurs. Elite character and readiness to die were thus documented. The doubling to ten squadrons ordered on September 24 was only accomplished the next spring. After that, eight squadrons were sent marching to Upper Silesia, where they served under the Dessauer. Two Hungarian squadrons stayed back in Berlin. The First Silesian War was over before it got to fire a shot. The peaceful pause was just the right time to firm up the regiment. It was expanded to ten squadrons, with 1172 soldiers, in Goldap and Stallupönen. In peacetime too, every hussar garrison had to set up an outpost every day outside the gate nearest the border and send out patrols to the nearest forest or village before the gates were opened or closed. For their main tasks took place in the field: reconnaissance, distant outposts to prevent surprise attacks, patrols to discourage desertion, surprise attacks to break up enemy forces, disturbing supply movement, attacking rear-guards, delaying marches, capturing small units, collecting contributions, securing marches as advance- and rear-guards. Mackerodt (1691-1743) had left Saxon cavalry service and joined Cuirassier Regiment 6 in 1717; he became a Cornet in 1718, a Major in 1738 and was raised to the nobility in the same year. Its first chief, for eighteen years, was Colonel von Ruesch, who—born in Hungary—came over as a Colonel in 1744 and was made a baron. Its most renowned chief was Daniel Friedrich von Lossow (1721-1783), a soldier since 1742, a lieutenant in Hussar Regiment 4 in 1748, a major in this regiment since 1759 and its chief after three years. A student of Seydlitz, he was an outstanding regimental commander and exemplary hussar leader at the age of 28, as well as a serious self-taught scholar of science and mathematics, French and military history, often visited by Immanuel Kant in Goldap. His officer corps profited from him, and in 1777 the King praised him "for his care in educating them so well". Replacements came from Dragoon Regiments 6, 7 and 8. Its garrisons from 1743 to 1763 were Goldap and Stallupönen, as of 1746 also Lyck, Johannisburg, Nordenburg, Lötzen, Pillkallen and Oletzko. Johannisburg, Nordenburg and Stallupönen were dropped in 1777, replaced by Schirwindt and Darkehmen. From 1779 to 1782 it was in Goldap, Ragnit, Tilsit and Stallupönen, and from 1783 to 1787 Ragnit and Tilsit were dropped and Lötzen, Stallupönen, Darkehmen, Rhein, Pillkallen and Bialla were added. As of 1792 Oletzko and Ragnit replaced Arys and Bialla. From 1796 to 1806 it was in Wirballen, Vistytis, Serrey, Przerosl, Wilkowischken, Suwalki, Kalwarija, Mariampol, Schirwindt and Prenn in Lithuania.

In 1744 it marched in the King's First Corps to Prague and then to southern Bohemia. In Zieten's rear-guard action at Moldauthein on October 9, it joined Hussar Regiment 2 in successfully driving back Ghilanyi's light troops. When the army's communication lines to Silesia were threatened, it endured the miserable retreat that began on November 22, in terrible weather, with the enemy never far away, "almost naked", as the King wrote about the regiment, "torn to pieces and completely wretched". In March of 1745 two squadrons went to mop up Upper Silesia under Hautcharmoy while the regiment was was in Steward Waldburg's Corps around Schweidnitz. On May 1 two squadrons under Winterfeldt beat Patacic's raiders at Hirschberg. On May 22, in the first battle line south of Landeshut under Winterfeldt, it broke forth against Nadasdy's superior forces and drove them over the Reichheh-hersdorf Mountains to Grüssau Abbey. On June 1 it went to camp at Stanowitz, joining du Moulin's Corps and leaving with them for Gräben on the evening of June 3. At Hohenfriedeberg it was on the right wing under Winterfeldt and struck Schlichting's Saxon cavalry southeast of Pilgramshain, driving them from the field and then attacking the grenadier corps. After that it secured the army on the upper Elbe. The uhlans who arrived in mid-July, also called "Bosniaks", were attached to the regiment after a short time with Hussar Regiment 2. In the attack on Kath. Hennersdorf on November 23 it joined with Hussar Regiment 2 and Cuirassier Regiments 8 and 9 to surprise the Saxon cuirassier regiments of Vitzthum, O'Byrn and Dallwig and the Saxe-Gotha Infantry Regiment, which were thoroughly beaten. It was allowed to keep the kettledrums of a cuirassier regiment. The pursuit to Bautzen and Zittau brought in prisoners, deserters and rich booty, leading to five days "without a single person complaining, but everybody accepting all difficulties with good will", as the King's secretary wrote. In Lehwaldt's Corps before Kesselsdorf it carried out reconnaissance and securing against Saxon uhlans as far as the Elbe. In peacetime it went to East Prussia; the weak "Bosniak" squadron went to Marggrabowa.

Since June 17, 1756 the King had been worned of Russia's arming, so that he ordered security measures for East Prussia, where the regiment was serving under Field Marshal von Lehwaldt. It was one of the weakest, numbering 36 officers, eighty non-commissioned officers, thirty trumpeters, smiths and medics, and 1080 privates. The russians appeared at the border at the end of June 1757. Lehwaldt drew back to the Pregel. At Gross Jägersdorf on August 30 it attacked Sybilski's cavalry in the Duke of Holstein's avant-garde and then on his right wing, beating them. In the forest battle there came the turn that led to an orderly withdrawal. When the Russians evacuated the

5. Husaren-Regiment

province all the same, the corps followed to Tilsit before the King called it back. In January and February of 1758 the Russians took East Prussia as far as the Vistula. In mid-February three squadrons were sent west to Duke Ferdinand's Allied Army. On February 23 they took 300 horses, eight standards and the kettledrums of the French Polleretzky Hussars at Stöckendrebber, not far from Nienburg. Then they advanced to Goch via Minden and Wesel. On July 23 they struck the French in the back at Krefeld and smashed them, but had to move back in July and were at Bork by the end of September. At the Rhine crossing they had wiped out a cuirassier regiment. The regiment fought in Dohna's Corps against the Russians in Pomerania, then in the Neumark. At Zorndorf on August 25, on the right wing under Schorlemmer, it led the attack against Demiku's mass action southwest of Zicher, which saved the infantry. After the battle it stayed in camp at Blumberg. In the fall it followed the Russians to Stargard, drove off a cavalry regiment at Greifenberg on October 26, under Platen, and relieved Kolberg. As of November 14 it took Eilenburg and advanced to Colditz. From late November to January of 1759 it fought in Near Pomerania, taking Demmin and Anklam and enclosing Stralsund. When it turned off to Landsberg via Stargard, one squadron stayed on the Peene under Kleist to oppose the Swedes, later moving to Torgau. The advance to Posen ended in Obornik. On July 23 at Kay, its six squadrons and Bosniaks under Wedell stormed the Russians on the Palzig Heights. During the Battle of Kunersdorf it stayed on the west bank of the Oder at Lebus under Wunsch to conceal the army's departure, take Frankfurt and prevent any move across the Oder. The squadrons in the Allied Army advanced on northern Hesse on April 13, attacked the French unsuccessfully at Bergen-Enkheim and had to retreat to the Weser. At Minden on August 1 they were able to defeat the French despite their possession of the fortress. At Gütersloh they shattered two French regiments.

In 1760 it served between Landsberg and Glogau and in the Hünern camp near Breslau in Prince Heinrich's Corps, which prevented the Russians from meeting the Austrians from mid-June to August 1, when the King arrived from Saxony. With the King on August 29, it served at Hohgiersdorf on September 17 and Guben in mid-October. When he went to Torgau, it returned to Silesia under von der Goltz to hold entral Silesia. Strengthened in 1761 to 41 officers, plus eleven non-commissioned officers and 150 privates for each of its seven squadrons, it guarded against the Russians at Glogau in May. In June it was ordered to Posen. Despite several very successful actions against superior light troops, the raid ended at Kosten. At Strehlen it joined the King and went to camp at Bunzelwitz. Under Platen it advanced via Gostyn and Landsberg to Freienwalde to hold off reinforcements and relieve Kolberg. After fighting its way into the city, it fought its way out toward the west two weeks later. While retreating to Alt-Damm it suffered losses at Schwenshagen. On November 14 it held Greifenberg again. At year's end it went to Leipzig, where its three transferred squadrons were formed anew. The length and duration of its marches were considerable. The Bosniaks were brought up to regimental strength. On the right wing at Burkersdorf in 1762 it led the feint toward Dittmansdorf and drove the outposts back to camp. At Reichenbach it attacked on the left flank under Lossow, complete with the Bosniaks, and captured three standards. Its eagerness to fight had brought exemplary achievements in small actions. After 1763 it served as an instructional regiment.

In 1806 it survived the catastrophe in L'Estocq's reserve corps and fought at Eylau and Heilsberg, for which all its officers received the Pour-le-merite. In 1914 it formed the Bodyguard Hussar Regiments 1 and 2.

5. Husaren-Regiment

HUSSAR REGIMENT 6

Commanders of the Regiment

1742	March	Colonel Isidor Count von Hoditz
1743	8/5	Colonel Carl Gustav von Soldan
1746	8/7	Colonel Ludwig Anton, Baron von Wechmar
1757	2/3	Colonel Johann Paul von Werner, later Lieutenant General
1785	1/30	Colonel Johann Benedikt von Gröling, later Major General
1791	12/29	Colonel Erich Magnus von Wolffrath, later Lieutenant General
1799	8/17	Colonel Ludwig Schimmelfennig von der Oye, later Major General
1807	1/25	Major General Friedrich Ferdinand, Prince of Anhalt-Köthen-Pless

In the autumn of 1741 King Friedrich II had the "Brown Hussar Regiment" organized in the unsecured province of Silesia and men recruited for it in and around Breslau and Ohlau. Formation began late in the fall; in January of 1742 two squadrons were established, in May it was in Ohlau, ready to march. The first squadron chiefs were Captain Piasedzki of Schirwindt, von Gersdorff, Captain von Jeanneret and Lieutenant von Netzow. The former Augstian Colonel, Count von Hoditz, of Rosswalde in Upper Silesia, became its chief in March but was discharged when peace was made on December 8 of the same year. It stayed in Silesia with Hussar Regiments 1, 3 and 4, stationed in Breslau. Its strength numbered thirty-six officers, 80 non-commissioned officers, 30 trumpeters, smiths and medics and 1020 hussars. On August 5, 1743 Colonel von Soldan became its chief. He came from Sweden and had been a Major in Hussar Regiment 1 since 1739. In training, foot drill was "minor activity" but "capable operation of the carbine" was required, so that the men could defend themselves in a local bivouac and "also force an enemy party that has taken a position in a churchyard or suitable place to move on foot". Pickets, sentries and patrols were supposed to secure a corps "like a spiders web, that one cannot touch without its being felt". In 1744 it joined von der Marwitz's Second Corps in upper Silesia, which did not participate in the march to Olmütz and advanced only to Jägerndorf and Troppau. Its eight squadrons waged a lively small-scale war with raiders. When it was attacked "by several thousand enemies" at Ratibor in December, it beat them despite heavy losses. On February 14, 1745 four of its squadrons attacked the Plomnitz Heights west of Habelschwerdt under Lehwaldt and drove the enemy back. On April 8 one squadron was taken prisoner with the Schaffstädt Detachment at Rosenberg. Then in April and May two squadrons under Hautcharmoy advanced to Gross Strehlitz, Konstadt and Kreuzburg. With Steward Waldburg's Corps near Schweidnitz since the end of April, two squadrons under Winterfeldt attacked the Patacic raiders at Hirschberg on May 1, rescued the magazine and took 130 prisoners. Under Winterfeldt again south of Landeshut on May 22, it not only fought off the attack of Nadasdy's superior forces but went from the first line to the counterattack and pursued Hungarians and Croats over the Reichhennersdorf Mountains to Grüssau Abbey. On June 1 it joined du Moulin's Corps at the Stanowitz camp, and on the eve of June 3 it reached its point of departure before Pilgramsheim, east of Striegau, for the Battle of Hohenfriedeberg. It encountered first the Saxon cavalry, then the cut-off grenadier corps. The pursuit reached to Landeshut. After a long series of small engagements in northern Bohemia its eight squadrons under Prince Dietrich arrived at Köthen to join the corps of Prince Leopold of Anhalt-Dessau, which secured against Saxony and stood at the border as of August 31. After the flank attack of Gessler and Kyau from the south at Kesselsdorf it attacked the Saxon left flank and rolled it up.

After Soldan's death, Colonel Baron von Wechmar became chief of the regiment, which had not been directed strictly enough, on August 7, 1746 and laid the groundwork for its later achievements. He came from Saxony and had joined the hussars as a Major in 1749, at the age of 28, and had been a Lieutenant Colonel since 1742. In 1757 he retired for health reasons. His well-known successor was Colonel von Werner (1707-1785), son of an Imperial hussar major in Raab; he became a soldier in 1723, Cornet in 1731, Captain in 1735, took part in 26 campaigns, including four against Prussia, joined Prussian service in 1750 and the regiment in 1751, and became its commander in 1756. The King recognized his accomplishments by promoting him to Major General ahead of sequence and to Lieutenant General in 1761, as well as with the Pour-le-merite; he valued him highly. His successor, of equal birth, was Colonel von Gröling (1736-1791), the son of a cuirassier, a private in Hussar Regiment 4 in 1745, chosen by Seydlitz to be a non-commissioned officer in his Cuirassier Regiment 8 in 1755, a Cornet in Hussar Regiment 5 in 1760, on Seydlitz's recommnendation a Major in Hussar Regiment 9 in 1762, in Hussar Regiment 5 in 1763, later in this regiment, ennobled in 1768, Commander in 1778. The regiment's replacements came from Cuirassier Regiments 4, 9 and 12, after 1798 from only 9 and 12. Its garrisons were Tarnowitz, Loslau, Beuthen O/S, Peiskretscham, Tost, Ujest and Gross Strehlitz from 1746 to 1753, with Gleiwitz, Rybnik, Sorau O/S, Nicolai and Berun replacing Tarnowitz, Tost and Gross Strehlitz since 1755. From 1764 it was in Pless, Peiskretscham, Nicolai, Beuthen O/S, Tarnowitz, Sorau O/S, Ujest, Katscher, Loslau and Rubnik from 1764 to 1777, plus Gleiwitz and minus Nicolai, Ujest, Katscher and Loslau until 1792. From 1796 to 1799 its garrisons were Gleiwitz, Beuthen O/S, Gross Strehlitz, Nicolai, Ujest, Peiskretscham, Pless, Loslau, Lublinitz and Zelasno from 1796 to 1799, with Rybnik instead of Zelasno from 1800 to 1806. From 1743 to 1756 the hussar regiments were directed in personal matters, training, economy and remounts by Colonel Hans Carl von Winterfeldt, who contributed significantly to their qualifications.

In 1756 it belonged to Schwerin's Silesian Corps, which led the diversionary attack over Nachod in mid-September. Attacked by 2000 cavalrymen as the avant-garde at Königgrätz, it drove them back and took 800 prisoners. When Lieutenant Colonel von Werner was sent behind the enemy lines with two squadrons by Schwerin, they beat heavily superior forces at Reichenau (32 kilometers east of Königgrätz and turned back almost without losses. In 1757 it took the Jungbunzlau magazine under Schwerin before standing on the battlefield at Prague on May 6. In Zieten's Corps on the left wing at the high point of the cavalry battle south of Sterbohol, it and Hussar Regiment 2 went over Hostiwarz to strike six regiments in the back and drive them to panic and flight. 1200 prisoners were taken, ten standards and the war chest were seized. At Kolin it was on the left wing in

6. Husaren-Regiment

Zieten's first line, preventing the cavalry troops of Serbelloni and Nadasdy from attacking southwest of the Eichenbusch. On the return march from Bohemia five squadrons fought their way through miraculously against fivefold superior forces at Gabel on July 15, but one squadron was captured in Gabel. In mid-August it went to Silesia under Bevern. On September 7 it went into battle at Moys, where Winterfeldt fell, then at Barschdorf on September 26; crossing the Oder south of Steinau, it took a position on the south bank of the Lohe at Breslau. Its eight squadrons under Zieten on the left wing withstood the enemy breakthrough at Kleinburg. Two squadrons were in Glatz. At Leuthen it belonged to the avant-garde and was in the center of the third battle line behind the infantry. Under Zieten it pursued as far as Landeshut. In 1758 it belonged to the King's Army with similar strength to Hussar Regiments 3 and 8: forty-one officers, ninety non-commissioned officers, thirty trumpeters, smiths and medics, and 1180 hussars, and took part in the taking of Schweidnitz and the enclosure of Olmütz. When the King went to the Oder on July 10, it stayed in Silesia under Margrave Carl, who went to join the King at Grossenhain, in Lower Lusatia, in September. Then it marched in Retzow's Corps from Bautzen to Weissenberg and arrived at the Hochkirch attack at the last minute. During the march to Silesia it led the cavalry battle at Ebersbach, near Görlitz, on October 26 and took 500 prisoners. On November 7 it helped to relieve Neisse, and two weeks later it was on the Elbe.

Strengthened to 100 non-commissioned officers and 1300 hussars in 1759, it went to camp with the King at Schmottseiffen, then under Prince Heinrich as of July 29, who came toward the King at Sagan after Kunersdorf and moved to Bautzen on September 5. Then it went via Hoyerswerda to Strehla and Torgau. In 1760 one battalion belonged to Prince Heinrich's Corps, the other to Fouqué's Cpros, which was beaten by greatly superior forces at Landeshut on June 23. The battalion fought its way through to Zieten with a cannon. The regiment was reuinted in Prince Heinrich's camp at Hünern. At Neumarkt it attacked two dragoon regiments and cut down those who were not taken prisoner. On September 6 it hurried with its chief via Landsberg and Schivelbein to relieve Kolberg with 800 men, strengthened by a squadron from Dragoon Regiment 5. At Sellnow on September 18 it broke through the Russian siege so remarkably that the siege corps left twenty-two guns and much material where it was and fled to their ships. Kolberg was saved and the Russian-Swedish fleet driven off by a few squadrons! The King gave two Pour-le-merite and had a medal struck, Werner's in gold. Then it tried to relieve Berlin in Württemberg's Corps. After Torgau it advanced to Köslin under Werner and secured the Rügenwalde-Schlawe line. When the Russians pushed it to the Rega, Werner reestablished the line. In 1761 it was the strongest regiment, with 120 non-commissioned officers and 1500 hussars, that was enclosed before Kolberg early in September. After breaking out, it was attacked at Treptow on September 12—Werner was captured—it reached Platen's approaching corps at Freienwalde and spent the unsuccessful end of the campaign with it. In 1762, now with 51 officers, it was in Silesia and successfully attacked at Reichenbach on August 16, under Lossow. In 1763 it had 1452 Prussians, 32 Saxons and 166 'foreigners'. In peacetime it served as an instructional regiment. Blücher's son Franz later became its commander.

In 1806 it fought in Hohenlohe's Corps at Jena and Zehdenick, fought its way through to East Prussia with 660 men, but lost three squadrons at Königsberg on June 14, 1807. Units went to Hussar Regiment 4, but the tradition was not recognized.

6. Husaren-Regiment

HUSSAR REGIMENT 7

Commanders of the Regiment

1744 4/12 Colonel Peter von Dieury, later Major General
1746 4/6 Colonel Heinrich Wilhelm von Billerbeck
1753 8/23 Colonel Paul Joseph Malachow von Malachowski, later Lieutenant General
1775 12/27 Colonel Adolf Detlof von Usedom, later Lieutenant General
1792 4/13 Colonel Fridrich Ludwig von der Trenck, later Major General
1796 6/24 Lieutenant General Georg Ludwig Egidius von Köhler, later General of the Cavalry, Governor in Warsaw

The formation of the regiment, already determined by King Friedrich II at the beginning of 1743, began on December 1 of that year in Köpenick, where Hussar Regiment 5 had left a squadron of Hungarians behind in 1742. Transfers from the six hussar regiments, such as seventeen men and horses per squadron from Hussar Regiment 1, filled it to its optimal strength. Volunteers from infantry and hussar regiments were rewarded with 5 Taler and sent to Köpenick. Mounted only in May and June of 1744, it went to Stolp, Lauenburg, Bütow and Belgard in Far Pomerania. Two squadrons stayed behind. Named as its chief on April 12, 1744 was Colonel Peter von Dieury, who came from Imperial service in 1743 as Lieutenant Colonel von Györy, became a Major General in 1744 and was discharged for health reasons in 1746. His successor von Billerbeck served until 1746 as a Lieutenant Colonel in Hussar Regiment 2, developed the peacetime training program and retired in 1753. Its Chief for 22 years was Paul Joseph Malachow von Malachowski, brother of the Chief of Hussar Regiment 3 from 1741 to 1745. He was already Chief as a Lieutenant Colonel after coming from Electoral Saxon service as a Captain, became a Major in Hussar Regiment 4 in 1745 and a Lieutenant Colonel in 1747. He repeatedly showed his bravery and became a Colonel in 1755, Major General in 1758 and Lieutenant General in 1771. His successor, Usedom, born in 1720, first served in Sweden, then in Austria, and since 1745 as a volunteer in Hussar Regiment 5, where he became Lieutenant in 1750, Captain in 1759, Major in 1763, Lieutenant Colonel in 1767, Colonel in 1772 and Commander in 1773.

The regiment's replacements came from Dragoon Regiments 9 and 10 until 1772, then from Dragoon Regiment 12. From 1746 to 1772 it was garrisoned in Soldau, Gilgenburg, Barten, Sensburg, Stappupönen, Bischofswerder, Rosenberg, Preussisch Eylau, Passenheim and Ortelsburg, then until 1788 in Schneidemühl, Bromberg, Inowraclaw (Hohensalza), Nakel, Chodziesen, Filehne, Schönlanke, Czarnikau, Lobsens and Uscz, as of 1787 also Schunin and Deutsch Krone. From 1789 to 1791 it was in Gniekowo, Inowraclaw, Barzin, Lojewo, Lobsens, Filehne, Parchany and Czarnikau, in 1792 in Schneidemühl, Nakel, Barzin, Schubin, Lobsens, Schönlanke, Deutsch Krone, Filehne and Inowraclaw. From 1796 to 1799 its garrisons were Kutno, Sagurowo, Kowal, Wartha, Klodawa, Kolo, Rawa and Ilow, in 1798 also Wolberzo and Szarek, from 1800 to 1806 Kutno, Kolo, Kowal, Konin, Stawiszyn, Piontek, Szadek, Klodawa, Uniewo and Slupca in Posen.

In the latter half of August 1744 its 1166 soldiers marched to Prague in the King's First Corps, then on to southeren Bohemia. Soon the enemy marched up. Raids and actions by the civilian population prevented reinforcement. On October 4 a unit leader Lieutenant Colonel von Eberstedt was smashed and its leader killed. On the march back after November 22 many Hungarians deserted, so that it was transfered to Züllichau for restoration after it returned. It was not exactly smart to apply it against its countrymen. In 1745 it was ordered to secure Upper Silesia north of the Oder against Hungarian raiders from April on: an independent small-scale war. In mid-August it went to Magdeburg under Prince Dietrich to join the corps of Prince Leopold of Anhalt-Dessau, which it reached at Wieskau on August 26. The march ahead turned out to be slow for many reasons: Leipzig on November 30, Eilenburg on December 3, Meissen on December 12. At Kesselsdorf on Decmeber 15 it and Hussar Regiment 6 covered the right flank of the right cavalry wing southwest of this key position, after it had driven off Sybilski's light cavalry as far as the edge of town. After the inclusive attack of the cuirassiers and dragoons, it rolled up the enemy position.

In peacetime it was transferred to East Prusaia under Lehwaldt as Commanding General. His quarters at Stolp were taken over by Hussar Regiment 8. The freedom of the cavalry leader in making decisions, changing formations and division of forces according to terrain and situation could be seen most clearly with the hussars. In the battle order, where it usually followed 200 paces behind the second line, they marched at a gallop to the outer wings "and swung squadron by squadron into the enemy's flank and rear, in order to beat him with the flank attack only recognizable at the end (hussar deployment). If necessary they could throw themselves against an enemy flank attack," as Jany writes. The original character of irregular troops had long since vanished. There were also 76 hussar officers from Hungary recorded, of whom five became generals, seven colonels, according to Jany. Not all of them stayed; many used the knowledge they gained in Prussia to qualify for foreign service. Expansion of the hussars and heavy losses forced the King to take them, though. In 1756 Russia's entry into the war was forseeable and thus danger to East Prussia, which was to be held. Only on January 22, 1757 did Russia ally itself with Austria, and the Russians appeared on the border at the end of June. The regiment had the same strength as the nearby Hussar Regiment 5. Lehwaldt's Corps was scarcely half as strong as the enemy, so he dropped back to the Pregel. In the Battle of Gross Jägersdorf on August 30 it was on the left wing under Lieutenant General von Schorlemmer and attacked the Russian cavalry stopped between the Norkitt Forest and the Pregel, penetrating deep into their position after orienting itself while marching up through the Puschdorf Forest to the north. The forest battle loosened control and made a retreat necessary, which the cavalry and artillery covered successfully. It followed the Russian retreat to Tilsit, where Lehwaldt received orders to march to Pomerania. When it arrived in Stettin at the beginning of December, the Swedes went back to Stralsund. In 1758 the King transferred two squadrons to Duke Ferdinand for the "Allied Army", which fell upon the French Polleretsky Hussar Regiment at Stöckendrebber on February 23 and was allowed to keep four standards and the kettledrums. After a successful advance to Goch they

7. Husaren-Regiment

could strike the French from behind at Krefeld and throw them out of their positions. At the Rhine crossing on the way back they smashed a cuirassier regiment. At the end of September they stopped on the Ems southwest of Münster. On December 20 Lieutenant Colonel von Narzynsky took command of the mixed battalion in the west, to be replaced at the beginning of 1761 by Lieutenant Colonel von Jeanneret. It had eight squadrons in the Pomeranian Corps that faced the Swedes in Near Pomerania until the Russians moved forward to the Warthe, then at Eberswalde and Frankfurt/Oder, and on July 17 in camp at Manschnow. At Zorndorf it rode under Lentulus in the Zabern Valley, taking part in Seydlitz's attack and circling the enemy's right flank along with Cuirassier Regiment 8 and Hussar Regiment 2. Then it flung itself at the infantry. Afterward it followed the Russians to Stargard from September on and was ordered from there to the Elbe. In November it attacked Colditz from Eilenburg. At year's end it went with the Hordt Free Regiment to Far Pomerania, soon to be followed by Platen with his dragoon regiment. In 1759 it was again with the Pomeranian Corps and advanced to Landsberg with seven squadrons until June 12. One squadron secured against the Swedes under Kleist. The advance along the Warthe was halted at Obornik as of June 24. In return, four squadrons crippled the Russian reinforcements to Bromberg. In the attack at Kay on July 23 it covered Züllichau under Wobersnow because additional manpower was thought to be in the Russian camp. In the afternoon it attacked but could save nothing. Before Kunersdorf it stayed at Lebus under Wunsch to disguise the army's departure to take Frankfurt and prevent any enemy withdrawal. Two squadrons took charge of protecting the Oder crossings. Then it secured against the Russians on the middle Oder while the one squadron from Peene went to the Elbe and the two in the west experienced defeat at Bergenand victory at Minden. In 1760 it joined Fouqué's Corps with eight squadrons, two of which went to join Zieten at Freiburg. Fallen upon by threefold superior forces at Landeshut on June 23, its chief was taken prisoner, but the greater part of it fought its way through very bravely. It was the regiment's day of honor! Zieten led those who remained to Breslau via Schweidnitz. Tauentzien could not use hussars in the fortress and sent them to Prince Heinrich's camp at Hnern. Under Goltz it secured northern Silesia. In 1761 it was strengthened to 41 officers, 88 non-commissioned officers and 1200 men and advanced to Posen but had to turn around in Kosten. Then it moved from the Bunzelwitz camp via Gostyn, Landsberg and Freienwalde to Pomerania under Platen, where it experienced much privation and ended up in Saxony. In 1762 it was brought up to ten squadrons with 110 non-commissioned officers and 1500 hussars to lead the King's action along the Schweidnitz foothills and help take the fortress.

In 1806 it fought under Rüchel at Jena, under Blücher at Lübeck and surrendered at Ratekau and Krempelsdorf on November 7, one squadron each at Lüneburg on November 11 and Boitzenburg on November 12. Three squadrons and the depot troops reached East Prussia and joined Hussar Regiment 4.

7. Husaren-Regiment

HUSSAR REGIMENT 8

Commanders of the Regiment

1743 11/17 Colonel Sigismund von Hallasz
1747 9/10 Colonel Alexander Gottlieb von Seydlitz, later Major General
1759 4/14 Colonel Otto Ernst von Gersdorff, later Major General
1761 1/3 Colonel Wilhelm Sebastian von Belling, since 3/5/1758 Commander of the new regiment, later Lieutenant General
1779 12/4 Colonel Carl August (von) Hohenstock, later of Hussar Regiment 5
1783 10/18 Colonel August Ferdinand von der Schulenburg, later Major General
1787 17/6 Colonel Johann Wilhelm, Count von der Goltz, died of wounds 8/13/1793
1794 6/4 Major General Gebhard Leberecht von Blücher, later Field Marshal

Its formation paralleled that of Hussar Regiment 7: though planned early in 1743, it began only as of December 1 of that year in Schwedt on the Oder. It was built around the second Hungarian squadron left behind in Köpenick by Hussar Regiment 5 in 1742, to which came transfers from the six hussar regiments. The remainder came from recruitment in specified infantry and hussar regiments, each man rewarded with five Taler. In addition to Schwedt, it was divided among Angermünde, Bahn, Greifenhagen and Freienwalde, on both sides of the Oder, where it was fully mounted and ready for service by June of 1744. As of November 17, 1743 its chief was Colonel von Hallasz, Hungarian by birth, who had just come over after fighting on the other side in the First Silesian War. In 1747 he was discharged because of age and weakness. His successor during the peacetime years was Colonel Alexander Gottlieb von Seydlitz, sometimes confused with the famous cavalry general. Following him was Colonel von Gersdorff, who came from Electoral Saxon service in 1741, was taken prisoner at Maxen and discharged in 1763 by the verdict of the Military Court. The place of the regiment, which was no longer capable of service as of November 21, 1759 and disbanded in 1763, was taken by the Belling Hussar Regiment, which had proved itself very well since 1758 as a free regiment and was now counted as the "Ninth" Hussar Regiment. It received replacements from Cuirassier Regiment 5 and Dragoon Regiments 1, 3, 4 and 5. It was garrisoned from 1746 to 1755—then taken over by Hussar Regiment 7—in Stolp, Schlawe, Bütow, Tempelburg, Lauenburg, Zanow, Neustettin, Rummelsburg and Belgard; as the Belling Hussar Regiment it was in Stolp, Lauenburg, Bütow, Schlawe, Zanow, Körlin, Bublitz and Tempelburg from 1764 to 1788, without Bublitz as of 1774, minus Bütow, Schlawe and Körlin in 1788, and from 1789 to 1791 Rummelsburg, Bütow, Neustettin and Schlawe returned. From 1796 to 1806 the First Battalion secured the border at Minden and Münster and was garrisoned in Münster as of 1802, the Second Battalion stayed in Bütow, Neustettin, Belgard and Rummelsburg, to which Stolp, Neustettin and Schlawe were added as of 1802.

In August of 1744 it had 36 officers, 80 non-commissioned officers, thirty trumpeters, smiths and medics, and 1020 hussars and belonged to the Second Corps of General von der Marwitz, which did not take part in the planned advance to Olmütz, so as to be able to protect Silesia. Until December it was stationed at Jägerndorf/Troppau, where it carried on a lively small-scale war. To fill vacancies in the army, Lieutenant Colonel von Weidenberg took a unit to recruit at the end of the year. On February 14, 1745 it was in Lehwaldt's Corps and successfully attacked the Plomnitz Heights east of Habelschwerdt across a snow-covered lowland crossed by brooks. At Hohenfriedeberg on June 4 of that year it had five squadrons in Zieten's cavalry reserve and did not come into action on the right wing. After that it went to Upper Silesia under Hautcharmoy, enclosed Kosel from the east on August 27, and advanced to Ratibor, where it distinguished itself at Kranowitz. The young regiment had proved itself and won two Pour-le-merite: Major von Schütz "for the coup at Steina" on February 1, and First Lieutenant Meerstedt for several actions around August 15, "who was beside himself with joy and could neither eat nor drink". It spent the peacetime on further training in Far Pomerania.

In 1756 it belonged to the Prince of Hesse-Darmstadt's reserve corps in Pomerania, which supported Lehwaldt in East Prussia. When the Russian threat diminished, the King moved the regiment to Sagan on October 19, and the corps followed to Lusatia by the end of December. In 1757 it had 41 officers, 90 non-commissioned officers, thirty trumpeters, smiths and medics and 1140 hussars, and was in Schwerin's Silesian Corps, with which it went to Prague via Jungbunzlau. There it was assigned to reconnaissance and securing around Böhmisch-Brod, from which Daun's Corps tried to unite with the main army. From the Kaiserweg at Kolin it covered the successful attack of Krosigk's Division at Krzeczhorz, which almost won a victory. When the army was divided at the end of the campaign, two squadrons went west to fight at Rossbach, while eight went to Silesia under Bevern. Marching via Moys to Barschdorf, Steinau and Breslau, it took up the firm position on the Lohe. Three squadrons took part under Zieten in fighting off the main Austrian attack at Kleinburg on November 22. At Leuthen it marched in the avant-garde, which drove the Saxon Chevauxlegers back beyond Gross Heidau, and then five squadrons stood in the center behind the infantry as the third battle line. After the battle it pursued as far as Landeshut, in accord with the orders of July 25, 1744 which said that the hussars were to pursue as far as possible and not allow the enemy any rest, even at night. In 1758 it went with the King's Army to Schweidnitz and Olmtz, the siege of which the King had to give up. When he marched to Zorndorf on August 10, it stayed in Silesia under Margrave Carl. As ordered on December 21, it was strengthened to 100 non-commissioned officers and 1260 hussars. In 1759 it was camped at Schmottseiffen, the command of which was taken over by Prince Heinrich on July 29. After Kuersdorf it went to Bautzen, arriving on September 12 and reinforcing Finck at Strehla on October 4. Previously it had been at Reppen, not far from Oschatz. Under Rebentisch on October 29 it smashed the threatened enclosure by Arenberg's Corps and received a Pour-le-merite. Under Finck it was moved forward by the King to the plateau of Maxen, where it was taken prisoner on November 21, including Major General von Gersdorff. No prisoners were exchanged until peace was made. As a result of the occupation of Far Pomerania, only three squadrons were formed in 1760, which were stationed in Breslau and other Silesian fortresses, where they remained in 1761. In 1762, with 41 officers, 90 non-commissioned officers and 1140 hussars, it belonged to the army in Silesia without taking part in the battles between Freiburg and Reichenbach. In 1763 it was disbanded as free troops; the reasons are unclear. The Belling Hussar Regiment took over its uniforms and garrisons.

8. Husaren-Regiment

At Prince Heinrich's request, the King authorized the establishment of a hussar battalion with five squadrons on February 16, 1758; it was to be in Halberstadt under Lieutenant Colonel von Belling and have 21 officers, 40 non-commissioned officers, fifteen trumpeters, smiths and medics, and 510 hussars. Two squadron chiefs were hussars, two from the cavalry, the lieutenants from the hussar regiments and the non-commissioned officers mainly from Hussar Regiment 1. The costs were met by the Hildesheim contribution. It grew so fast that by mid-May it had 140 supernumeraries, but doubling it was denied. Belling (1719-1779) had been in Hussar Regiments 1, 2 and 6 since 1739 and a squadron chief since 1747. A student of Zieten and one of the best hussar leaders, he became the gifted teacher of Günther, Goeckingk, Blücher and Kalckreuth. In May and June it was already raiding in Franconia; then it was stationed at Gross Sedlitz and across the Müglitz, where it just avoided being surrounded. In an attack on the Freiberger Vorstadt it struck down many and took 200 prisoners. On April 15, 1759 Belling used two squadrons and 360 riders to smash three battalions at Sebastiansberg, taking over 1300 prisoners, three cannon and three flags. In May it attacked in Franconia and the Upper Palatinate with the Saxon Corps and reached the Main. On July 30 it went with the King from Sagan to the Oder. In the first battle line on the left wing under Seydlitz at Kunersdorf, it flung itself against the advancing enemy cavalry on the Kuhberg and chased them to the Deep Path without being able to spread out. After securing northern Silesia against the Russians, it went with Dragoon Regiment 3 and Hordt's Free Regiment under Manteuffel to oppose the Swedes at Prenzlau in mid-September. In successful small actions it pushed them back to the Peene. With Dragoon Regiment 7 in Stutterheim's Corps in 1760 it fought successfully during the withdrawal to the Ucker, for example, at Jagow, where it took 200 prisoners. When the corps relieved Berlin, it stayed at Templin with Hordt's Regiment to oppose the Swedes and capably held them in check, and returned to the Peene in October. On January 3, 1761 it gained its second battalion in Mecklenburg, and its third in Saxony on account of a crowd of volunteers, the latter going to Hussar Regiment 6 in Far Pomerania. The regiment stayed in Near Pomerania all year, boldly and enterprisingly keeping the Swedes on edge. In 1762, with 41 officers, 120 non-commissioned officers, thirty trumpeters, smiths and medics, and 1460 hussars, it joined the army in Saxony, where the Third Battalion rejoined it. After the breakthrough over the Mulde on May 12, it pushed the Imperial army back to Münchberg, southwest of Hof, under Seydlitz. Then it advanced into Bohemia to destroy the magazines at Eger, Elbe and Leitmeritz, collect contributions and take prisoners, and fight at Teplitz on August 2. After that it went raiding as far as Bayreuth and through northwestern Bohemia. At Freiberg it secured the Struthwald in Stutterheim's column, attacked Klein-Walterdorf and surrounded the enemy, taking many prisoners and a flag. Decreased to ten squadrons in 1763, it had 610 Prussians, 124 Saxons and 487 'foreigners'.

After great success in the 1793-1794 campaign, it was called "Blücher's Hussars" as of March 20, 1794. In 1806 it fought at Jena and avoided surrender at Ratekau on November 7, six squadrons going to East Prussia, later becoming Hussar Regiment 5 and maintaining the tradition since 1887.

Husaren-Rgt. Belling v.,
später Husaren-Regiment Nr. 8

BOSNIAK REGIMENT 9

Commanders of the Regiment

1762 3/17 Colonel Daniel Friedrich von Lossow, Commander of Hussar Regiment 5
1783 10/18 Colonel Carl August (von) Hohenstock, Commander of Hussar Regiment 5
1788 2/3 Colonel Heinrich Johann von Günther, later Lieutenant General
1803 4/30 Major General Anton Wilhelm von L'Estocq, formerly Chief of Hussar Regiment 1

The King's only completely foreign troop came about by chance. In the summer of 1745, when Polish-Saxon uhlans tried to cross the Oder into Saxony at Schiedlo, north of Guben, under Major General von Bardeleben, the Bosniak Regiment of Colonel Ignatius Manzani von Slavedin was left by half a company under Captain Serkis, with two officers, two non-commissioned officers and 45 men. They reported to Major General Dieury of Hussar Regiment 7 in Züllichau on July 17 to join Prussian service. Marching via Glogau, they reached Breslau on July 26. As of August 1 they were in camp at Chlum, earning 308 Taler, 18 Groschen a month in the King's employ, at first with Hussar Regiment 2, then Hussar Regiment 5 until they became independent. Prussian tolerance had no problems with Moslems. As early as September 26 they smashed a raiding party in Greiffenberg, "whereby the Bosniaks, mixed in among the regular Hussars, behaved uncommonly bravely". They fought under Captain von Warnery of Hussar Regiment 4, reorganized into a hussar regiment only two years before. For the Balkan riders this test of their ability with the lance, as a "new weapon", was successful. On November 23 they took part in the attack on Kath. Hennersdorf, and mopped up on the middle Elbe in mid-December. In February of 1746 there were still four officers, four non-commissioned officers and 35 privates, soon strengthened by the Saxon-Polish Captain von Krzeczewsky with eighteen privates to five non-commissioned officers and 53 privates. From 1748 to December of 1755 its strength sank to one officer, four non-commissioned officers and 39 privates. The budget always listed an optimal strength of five officers, six non-commissioned officers and eighty men, and paid for them until 1761. Every private received five Taler, 21 Groschen as his salary plus rations. Out of this clothing, weapons and horse had to be maintained. For years the small corps did not wear Prussian uniforms, and it obviously knew how to keep the King interested in it. There was no regular replacement; it depended on volunteers and recruits. In 1754 its garrison was moved from Marggrabowa to Oletzko, in 1755 it was in Pillkallen, as of 1764 in Goldap, Stallupnen, Bialla and Arys, from 1770 to 1792 in Ragnit, Johannisburg, Nikolaiken, Lyck, Rhein, Stappupönen, Bialla, Passenheim and Sensburg. In 1796 it was sent out to Lomza, Ostrolenka, Mlawa, Zakryczym, Kleszczle, Prasznycz, Bransk, Wyzkowo, Ostrow and Knyszyn in southern Prussia, and from 1796 to 1806 it was in Tykocin, Ostrolenka, Drohyczin, Knyszyn, Zabludow, Bransk, Wyzkowo, Ostrow and Lomza. Until 1788 it had the same chiefs as Hussar Regiment 5, among whom Daniel Friedrich von Lossow (1721-1783), a student of Seydlitz, stood out. Then from 1788 to 1803 it had the militarily highly gifted, manly, spirited, widely educated Heinrich Johann von Günther (1736-1803), a baron as of 1798, son of an army chaplain, a soldier since 1756, determined, self-assured, as well as goodhearted, Zieten's and Belling's student, who earned great honors.

Service and action in the Seven Years' War ran almost parallel to those of Hussar Regiment 5. In 1756 it was in East Prussia under Field Marshal von Lehwaldt since mid-July, ready along with three hussar and five dragoon regiments. At Gross Jägersdorf on August 30 it proved itself against the Cossacks on the right wing, then marched to the mouth of the Oder. In 1758 it fought against the Russians in Pomerania, the Neumark and at Zorndorf on August 25, where it met the Russian flank attack on the right wing. Then it helped to relieve Kolberg, broke through the position on the Mulde and stayed on the Peene in Near Pomerania until the start of 1759. After that it went after the Russians in the Posen area and at Kay on July 23, then secured the west bank of the Oder at Kunersdorf. In 1760 it held the Russians in check between Landsberg and Breslau. In May of 1761 Colonel von Lossow, at the urging of Lieutenant General von der Goltz, had the Bosniaks, which had decreased to one officer, five non-commissioned officers and sixteen privates, expanded to 110 horses through recruitment, after a rumor of an alliance with Turkey had lured Turkish and Tartar turncoats. Their usefulness and stature had results. Strengthening to a full regiment followed in 1762. It was again used to oppose the Russians in Posen and Far Pomerania. The recruitment of eight companies of 100 riders each, beginning on November 11 with the cooperation of the Polish Colonel Florian Murza Krzeczowsky did not come to pass. But Lossow expanded the corps on March 17, 1762 to ten squadrons with forty officers, a hundred non-commissioned officers, thirty trumpeters, smiths and medics and 830 privates, making it a full regiment of 1000 soldiers. It was with the army in Silesia from the beginning of July, serving at Kanth, Hohenfriedeberg and Adelsbach, at Burkersdorf, where it drove the enemy back from Dittmannsdorf through Reussendorf, and Langenbielau, where it attacked successfully under Lossow. Then one battalion covered the siege of Schweidnitz, while the other went to Saxony and arried the small-scale war into the neighboring areas. After peace was made it was reduced, despite its good record, to two squadrons with five officers, sixteen non-commissioned officers, four trumpeters, smiths and medics and 200 privates and turned over to Hussar Regiment 5; the majority of its men were transferred to other regiments.

On June 1, 1770 the King had the corps expanded to five squadrons, each with five officers, eight non-commissioned officers, three trumpeters, smiths and medics, and 100 privates, a total of 580 soldiers. Because of the planned doubling, it was given a full officer corps in 1771 and divided into ten squadrons of fifty privates each. As ordered on December 6, 1771, another 500 men were to be recuited in 1772 but remain unmounted until 1773. As of June 1, 1772 the budget included fifty officers, 100 non-commissioned officers, fifteen trumpeters, smiths and medics, 100 Prussians and 900 foreigners. In 1772 the regiment gained the King's full recognition in the review. By cabinet order

9. Husaren - Regiment
Bosniaken (Uniform um 1780)

of April 13, 1773 it reached the increased budgeted figures of fifty officers, one hundred ten non-commissioned officers, thirty trumpeters, smiths and medics, a hundred Prussians and 1220 foreigners. But provision with mounts was not to be completed until 1775. It filled its ranks through its own recruiting. The Chief of Hussar Regiment 5 remained its chief. In the 1778 mobilization another squadron was formed under Captain Moczydtosky. The increased Polish population and the fragmenting of the Polish army certainly helped the regiment grow. In the War of the Bavarian Succession it was in the King's Army in northern Bohemia. As of January 4, 1779 Lieutenant Colonel von Schill, the hero's father, began to form a battalion of Tartars with five squadrons in Soldau, Willenberg and Gumbinnen; because peace was made, he had to give it up. Fifty mounted Tartars thus joined the regiment. On February 3, 1788 it received its own Chief, Colonel von Günther, and became independent, though still linked with Hussar Regiment 5. Mobilized on April 24, 1794 because of the uprising in Poland, it came to the Narew under Günther and fought off attacks at Piontnica on June 25, put up a renowned fight at Kolno on July 10, and drove a Polish corps back over the river and captured the battlements with all the guns at Demniki on July 18. On October 26 Lieutenant Colonel Schimmelfennig von der Oye fell on the enemy's rear with three squadrons at Gross Magnuszewo and took more than 400 prisoners and five guns, after two squadrons had chased General von Kornowski's forces across the Narew at Rozan on October 8. At the request of Colonel Janos Murza Baranowski, the King granted the Tartars of new East Prussia not only free practice of religion and a free dwelling area, but also a corps of light cavalry, ordered on October 8 as a battalion of five squadrons in the regiment. There were 21 officers, ten sub-ensigns, thirty non-commissioned officers, 21 trumpeters, smiths and medics, 250 Tartars of the lower nobility, called "Towarczys", 250 Christian privates, six members of the staff, a total of 528 soldiers, though they never reached full strength. The name comes from "towarzysz": comrade. They formed the first unit, obeyed only their officers and sub-ensigns, brought their own weapons and uniforms and were free from dishonoring punishments. They served for twelve years and were paid six Taler per month, the privates ("Pacholken") were in the second unit, served twenty years and were paid half as much. Polish uniforms were still being used. The regiment was in charge of training, also involving non-commissioned officers of Hussar Regiments 5 and 10 because of the Hussar Regulations being translated into Polish. In 1795 it was stationed in Augustowo, from 1796 to 1806 in Augustowo, Goniondz, Suchowola, Rajgrod and Janow, plus Sokolka as of 1801, and with Lipsk instead of Goniondz as of 1804. Until 1806 only the fifth squadron in Rajgrod remained Tataric.

After the end of Poland on October 24, 1795 the King ordered the reorganization of the regiment and its Tartar battalion on October 14, 1799, to consist as of June 1, 1800 of ifteen squadrons of Towarczys with 51 officers, 100 non-commissioned officers, forty trumpeters, smiths and medics, 1080 Towarczys and a five-man lower staff for the regiment and 26 officers plus half of everything else for the Towarczys' Battalion. At this point the Prussians were transferred to Hussar Regiment 5 in exchange for Towarczys; thus it was made thoroughly Polish. The regimental staff was stationed in Tykoczyn, that of the battalion in Augustowo. On May 3, 1800 the Towarczys were confirmed as a fourth type of cavalry through the establishment of an inspection.

In 1806 it belonged to its chief L'Estocq's reserve corps and fought bravely at Preussisch Eylau and Heilsberg. On July 26, 1807 it became an uhlan regiment, later Uhlan Regiments 1 and 2.

9. Husaren-Regiment
Bosniaken

HUSSAR REGIMENT 10

Commanders of the Regiment

1773	8/1	Colonel Carl Christoph von Owstien
1780	6/15	Colonel Friedrich Wilhelm von Wuthenow, later Major General
1787	3/23	Major General Friedrich Gideon von Wolky, later Lieutenant General
1797	1/23	Colonel Carl Ludwig von Lediwary, later Major General, formerly Chief of Hussar Regiment 4
1799	10/1	Colonel Friedrich Daniel von Glaser, later Major General
1804	12/3	Colonel Friedrich von Usedom, later Major General

Within the framework of military organization in the new West Prussian districts, King Friedrich II had already planned as of October 6, 1771 not only four infantry regiments but also another hussar regiment of 800 men. Securing the open eastern border really required light cavalry. By cabinet order of March 3, 1773 it appeared in the budget as of June 1. The formation of ten squadrons began in July at the former garrisons of Hussar Regiment 7 in Soldau, Ortelsburg, Gilgenburg and the small border towns on the Drewenz after Hussar Regiment 7 had been transferred to the Netze area. Officers and a cadre of men came from the existing hussar regiments; some non-commissioned officers and sergeant-majors were also transferred to be lieutenants and first lieutenants. The equipment depot, which had been moved from Breslau to Stettin in 1771 as a precaution, was now also moved to Soldau. The enlisting of foreigners came to a stop at the end of the year; on April 15, 1774, canton soldiers followed for two months of training. As of 1782 it was given the district of Bromberg as its canton, plus the administrative areas of Koronowa and Bartelsee and the cities of Bromberg, Koronowa, Schulitz and Fordon. The number of horses grew to 1000 in 1774; the optimal number of 1440 was reached only in 1775. On August 1, 1773 Colonel Carl Christoph von Owstien became its chief. He was succeeded in 1780 by Colonel von Wuthenow (1723-1801), a Colonel since 1775 but not very successful. Next came Major General von Wolky (1735-1803), a Colonel since 1785, who proved himself in 1794 at Warsaw and on the Narew. Colonel von Lediwary (1735-1812) became Chief in 1797, after being Chief of Hussar Regiment 4 since 1794. Colonel von Glaser (1741-1804) was formerly Commander of the Bosniak Regiment and was given Hussar Regiment 10 when the Bosniaks were reorganized into the "Towarczys' Corps". Colonel von Usedom (1757-1824) had spent his career in Hussar Regiment 5 and become a Colonel in 1798. From 1773 to 1795 the regiment was garrisoned in Soldau, Passenheim, Gilgenburg, Strasburg/West Prussia, Löbau, Neidenburg, Ortelsburg and Neumark/West Prussia; in 1796, after the uprising of the Republican Polish Army, in Warsaw and Gora; after Poland ceased to be, from 1797 to 1799, in Warsaw, Gora, Nowy, Dwor, Biezun, Lipno, Mszczonow, Rypin, Praga and Nowe Miasto, and in 1799 Skiernewice replaced Gora. From 1800 to 1806 the cities of Praga and Nowe Miasto were dropped in favor of Blonie, Rawa Mazowiecka and Raczions. The acquisition of West Prussia with the Netze district had markedly improved the possibilities for defending East Prussia. The King called himself "King of Prussia from then on. A Pole could not be a German but could certainly be a good Prussian!

The rebuilding of the battered army was the main job of the peacetime years. The quality of the officer corps had to be raised again. Training and education needed improvement. Instruction and drill were supposed to maintain the fighting strength in the field. The question was whether the former spirit of the army could be preserved. In view of the lack of recruits, the expansion of the army had to depend more strongly on 'foreigners', so that they made up half the army by 1786. The difficulty of maintaining discipline did not decrease. The harshest punishment for the hussars was transfer to the infantry. Every regiment had to pass the spring review, which included both regimental inspection and drill, at which the King noticed the smallest details. Along with the general review of a provincial corps, there were also school maneuvers with two parties. In the autumn, or sooner in case of emergency, there came the great autumn maneuvers with movement over a large area. When the War of the Bavarian Succession began in 1778 after only a few years, it belonged to the Second Army of Prince Heinrich, which gathered on both sides of Berlin between the Elbe and Oder and also included Saxon troops. Early in July it marched over the Elbe into northwestern Bohemia to force the Austrians out of their position on the Iser. The army reached the line formed by Reichenberg, Niemes, Böhmisch Leipa, Leitmeritz and Melnik. But Prince Heinrich delayed and began the march back to Zittau and Aussig on September 10. He had avoided every battle. He also cited lack of fodder and epidemic illness. The King complained that the enterprising spirit of the leadership was what was lacking, "real hussar service has disappeared almost completely". Actually, the troops had really done their duty in all actions. In 1787 it was enlarged to 51 officers, 150 non-commissioned officers, thirty trumpeters and 1320 hussars. In case of war, a "depot" of six officers, twenty non-commissioned officers and 160 men was to be prepared as a unit of replacement troops.

When it was mobilized under Usedom at Bromberg-Hohensalza in November of 1788, the troop was reminded to treat the well-intentioned Poles kindly. In June of 1789 it was in Usedom's army at Hohensalza/Soldau "to protect Poland from external pressure". As of July 5 it marched southward straight through poland to Lublinitz O/S. After the danger of war had passed at the end of July, it turned back to West Prussia as of August 3, but was demobilized only in August of 1791. Only a quarter of the army was involved in the 1792-1795 campaign against France; the hussars proved themselves particularly well there. The mass of the army held the Polish border, as the Russians had marched into Poland in May of 1792, but only in September of 1793 was it mobilized, as the second partition of Poland dragged on. Five squadrons secured east of the Vistula between Soldau and Plock. The uprising of the republican army began with the attack of tthe Polish National Cavalry Brigade of Madalinski on the unit of Lieutenant Colonel von Tümpling, composed of eight non-commissioned officers and 49 hussars, at Srensk, south of Soldau, on March 15, 1794. After a brave fight, Tümpling was captured with most of his men; the others escaped to Soldau. The Poles crossed the Vistula. The Chief of the regiment, General von Wolky, took on the remaining Russian occupation troops, under General Baron Igelström, on the western edge of Warsaw on April 18, with several squadrons

10. Husaren-Regiment

involved. Despite the seriousness of the situation, the army at first limited itself to securing the border, and mobilized 40,000 men only on April 24. From late May on, under its chief, it secured about eighty kilometers of the west banks of the Bug and Narew between Zakroczym and Rozan, an area of many woodlands and swamps and crossed by numerous waterways, thus difficult to oversee. In abrasive small actions against individual units, often under attack from the rear as well, it defended this sector with a few battalions "with greatest bravery" for six months against heavily superior enemy forces, without at first getting into large-scale battles. Exposed to constant small-arms and cannon fire, it tried to stop the enemy at river crossings and prevent it from breaking through to the west. During the attack on Warsaw it kept active: on August 18 and 24 it came tonumerous river crossings at Segrze, Serock and Dembe in the bend of the Bug, but was driven back at all points. When General Dombrowski crossed the Vistula with 2000 men at Schülitz for a surprise attack on Thorn, the Regimental Commander, Colonel von Lediwary, drove him back successfully with four squadrons and four battalions and took prisoners and guns. Although the capture of Kosciuszko on October 10 broke the back of the uprising, the Poles still tried to seize the Orzyc crossing at Gross Magnuszewo with 6000 men on October 10. Four squadrons with several companies and three Bosniak squadrons in the rear smashed the enemy and took 400 prisoners and five guns. At Bromberg one squadron helped to pursue remaining enemy troops as far as the Bzura. The regiment had passed its first test.

The First Battalion of the regiment stayed in southern Prussia, the Second Battalion in new East Prussia, where the troops' lodgings were pitiful and improvements could be made only gradually. In 1806 it fought in Württemberg's Corps and Blücher's Second Corps at Halle, took part in the march to Lübeck but surrendered at Wismar on November 5. Seventy-five hussars were taken prisoner at Küstrin's capitulation on November 1, and one detachment had to surrender its weapons at Magdeburg on November 11. Two squadrons, the depot and the baggage train escaped to East Prussia and formed the Third Squadron of the First Hussar Brigade of Colonel von Corvin-Wiersbitski in 1807. Later it joined Hussar Regiment 4.

Since 1763 the hussar commands of Prince Heinrich still existed in Magdeburg and Rheinsberg, as well as the one that had been in Berlin since 1792 after the takeover of Hussar Battalion 11, established in Ansbach-Bayreuth under von Franckenberg, von Saas and von Bila.

10. Husaren-Regiment

REGISTER OF NAMES

WHY AND HOW TO WAGE WAR
PRINCIPLES OF WAR

Battles decide the fate of a state. When one wages war, one must dare to take decisive steps, either to draw oneself out of the difficulties of war or to push the enemy into them, or to end disagreements that otherwise would never end.

A rational man must never take a step without having reason to. Much less must the general of an army do battle without seeking to attain an important purpose thereby. If he is forced into battle by the enemy, then he has surely stepped into a difficult situation that compels him to accept his enemy's conditions.

One will see that I am not speaking in praise of myself. For among the five battles which my troops have had with the enemy there have not been more than three that I had decided on in advance, and I was forced into the other two: that at Mollwitz, because the Austrians placed themselves between my army and Ohlau, where my artillery and my provisions were, and that of Soor, because the Austrians cut my path to Trautenau and I could not avoid a battle without being ruined. But one sees what a difference there is between forced battles and previously planned ones. What kind of success would not the battles of Hohenfriedberg and Kesselsdorf and that of Czaslau have had, which brought on peace.

When I give instructions as to how one is to conduct battles, I cannot deny that I have often disobeyed them. But my officers are to derive benefit from my mistakes and know that I shall do all that I can to improve.

The best battles are those which one forces the enemy to fight. For it is a proved rule that one must compel the enemy to do what he has no desire to do, and because my interests oppose those of the enemy, thus I must want that which the enemy does not want.

The causes because of which one must do battle are as follows:
1. to force the enemy to raise the siege of a place;
2. to drive him out of a province that he has taken by force;
3. to penetrate into his land;
4. to undertake a siege;
5. to break his will when he does not want to make peace; or
6. To punish him for a mistake.

One compels the enemy to strike when one makes a forced march, whereby one gets behind him and cuts him off from his supply lines, or when one threatens a city which it is important for him to keep.

But one must be careful when one makes such a maneuver, and be careful that one does not put oneself in the same position or position oneself so that the enemy can cut us off from our magazines.

In attacking the rear-guard one risks the least. When one wants that, one lurks very close to the enemy, and when he wants to withdraw, one falls upon his rear. In such an attack one risks little and wins much.

Small skirmishes are meant to prevent the attack of an enemy corps. A capable enemy will escape easily in a forced march or be able to take a position that he considers good.

At times one is not willing to let oneself go into battle, but is often lured into it by a mistake of the enemy's and is obligated to utilize it in order to punish him for it.

To all these maxims I add that wars must be short and lively. Our advantage does not allow them to be drawn out. A lasting war weakens our splendid manpower, depopulates the land and exhausts our sources of help.

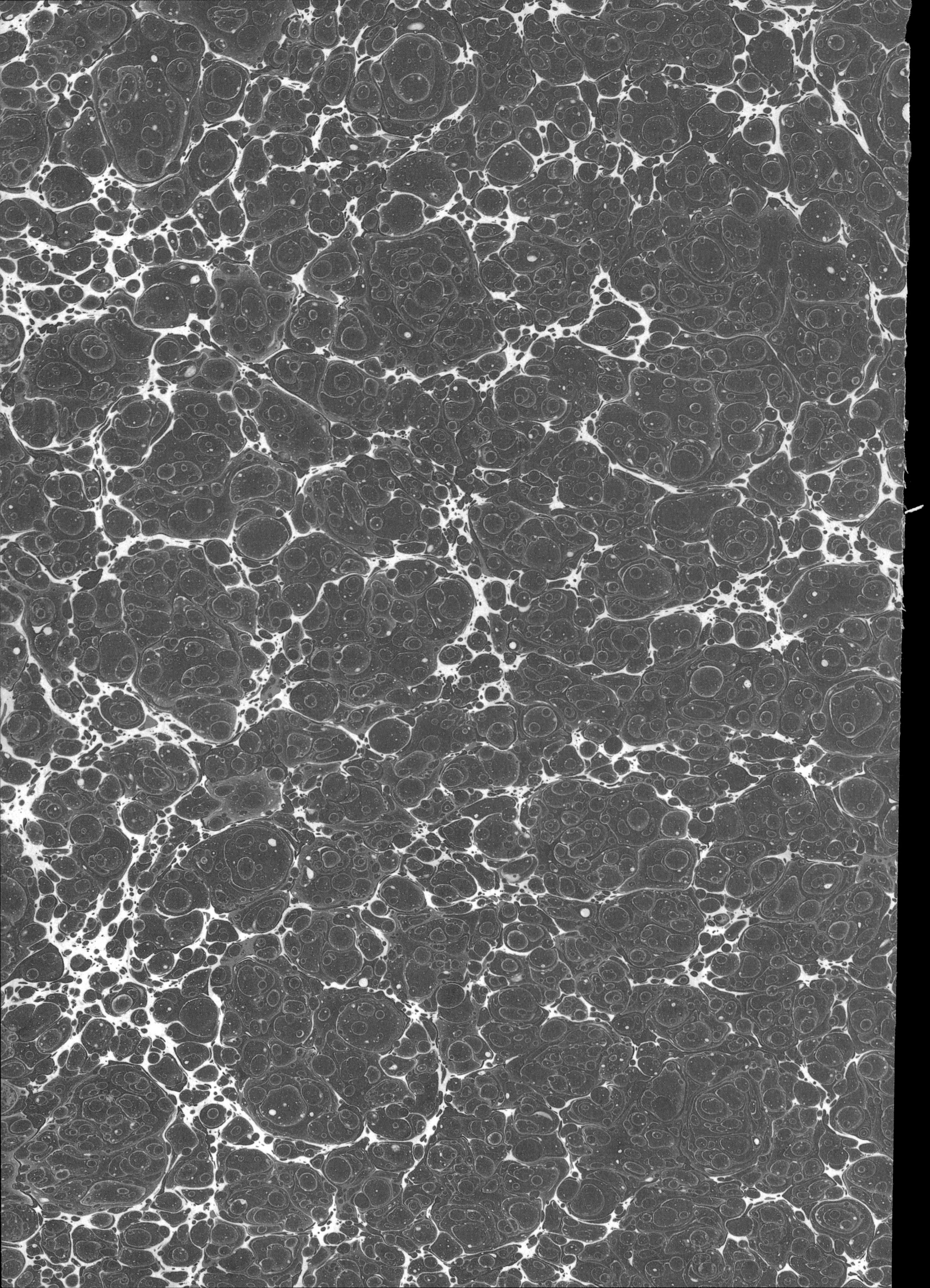